Cognition-based Smart Computing

The book guides the smart computing community toward future directions in cognitive research and demonstrates how these advances can be applied to next-generation smart systems. It further identifies applications that demonstrate innovative ways of utilizing cognition-based smart computing for solving various real-world problems.

- Presents cognitive modeling in machine consciousness and human cognition.
- Covers artificial intelligence (AI) and its applications in machine self-learning, human-computer interaction, natural language (NL) processing, cognitive models, and architectures.
- Discusses machine learning (ML) and cognitive computing for mobile communications, and wireless networks.
- Explains the foundation, framework, and applications of cognition-based smart computing in a single volume.
- Showcases key issues and challenges in the emerging area of cognitive computation and promotes an interdisciplinary understanding of the diversity in the adaptation of cognition in smart computing.

The text is primarily written for senior undergraduates, graduate students, and academic researchers in the fields of electrical engineering, electronics and communications engineering, computer science and engineering, and information technology.

Cognition-based Smart Computing

Foundations, Frameworks, and Applications

Edited by

Sangita S. Chaudhari, Pallavi Vijay Chavan, Parikshit Mahalle, and Pranav Pawar

CRC Press
Taylor & Francis Group
Boca Raton London New York

CRC Press is an imprint of the
Taylor & Francis Group, an **informa** business

Designed cover image: Shutterstock

First edition published 2026
by CRC Press
2385 NW Executive Center Drive, Suite 320, Boca Raton FL 33431

and by CRC Press
4 Park Square, Milton Park, Abingdon, Oxon, OX14 4RN

CRC Press is an imprint of Taylor & Francis Group, LLC

ISBN: 9781032795331 (hbk)
ISBN: 9781032817651 (pbk)
ISBN: 9781003501220 (ebk)

DOI: 10.1201/9781003501220

Typeset in Sabon
by Newgen Publishing UK

Publisher's note: This book has been prepared from camera-ready copy provided by the authors.

Contents

Preface

Smart and cognition-based computing has captured the attention of virtually every industry vertical. With the enormous growth of data generated each day, we continue to equip ourselves with advanced tools and technologies to manage and interpret this data through smart computing. Cognition-driven smart computing has become essential across domains, especially in the industrial sector. Many e-commerce businesses rely heavily on intelligent decision-making, collecting massive amounts of data daily and using it to shape policies that improve their operations. At this stage, the role of smart computing becomes indispensable.

Smart and cognition-based computing has captured the attention of virtually every industry vertical. With the enormous growth of data generated each day, we continue to equip ourselves with advanced tools and technologies to manage and interpret this data through smart computing. Cognition-driven smart computing has become essential across domains, especially in the industrial sector. Many e-commerce businesses rely heavily on intelligent decision-making, collecting massive amounts of data daily and using it to shape policies that improve their operations. At this stage, the role of smart computing becomes indispensable. Artificial intelligence (AI) contributes significantly to the development of intelligent applications in data science, particularly through the design of rule-based systems that incorporate facts and inference rules. Additionally, soft computing, fuzzy logic, and genetic algorithms offer valuable approaches that further enhance smart application development.

Today, intelligent applications are used across nearly every sector, including healthcare, education, agriculture, finance, gaming, sports, and entertainment. In healthcare, such applications support diabetic retinopathy analysis, diabetes detection, and expert systems for medical diagnosis, among others. In agriculture, popular intelligent applications include rainfall prediction, weather forecasting, crop prediction, soil quality assessment, and more. This book aims to provide a platform for researchers to advance intelligent applications and their design for the benefit of technology and society as a whole.

This publication comprises twelve chapters, beginning with foundational concepts in cognition-based smart computing and progressing through a range of tools, methods, and techniques central to the field. **Chapter 1** explores AI-driven optimization for multi-objective task scheduling in fog-assisted IoT environments. **Chapter 2** examines various approaches to transforming healthcare through cognitive and smart computing. **Chapter 3** focuses on cognitive cybersecurity supported by symbolic deep learning (DL)–enhanced decision systems and discusses the use of pre-trained deep learning models. **Chapter 4** investigates the application of the Dream Sleep algorithm to multi-layer perceptrons to mitigate catastrophic interference, emphasizing the role of artificial neural networks. **Chapter 5** presents an integrated approach to threat intelligence for cognitive ransomware detection in 5G networks, situated within the broader security domain. **Chapter 6** provides a comprehensive study of generative AI algorithms and techniques relevant to cognitive computation. **Chapter 7** discusses the enhancement of object recognition through the fusion of YOLO and OCR, outlining a roadmap toward more intelligent cognitive tasks. **Chapter 8** addresses cognitive robots for Industry 5.0, covering fundamental concepts, existing challenges, and emerging research trends. **Chapter 9** highlights the role of cognitive technologies in precision agriculture within the Industry 5.0 paradigm, focusing on the integration of recommendation and prediction systems to support transparent farming transactions. **Chapter 10** examines applications of cognitive computing in the banking and finance sectors. **Chapter 11** investigates the role and potential of large language models in advancing cognitive computing. Finally, **Chapter 12** provides an overview of electroencephalogram feature extraction methods for stress analysis.

We would like to express our sincere thanks and gratitude to the authors for their outstanding contributions to smart cognitive computing. Without their efforts, preparing and submitting this proposal would have been extremely difficult. As editors, we hope that this book will inspire and encourage researchers to pursue further work in this domain. We also expect that researchers will benefit from the wide range of applications presented throughout this book. Our special thanks go to our publisher, CRC Press/Taylor & Francis Group.

We hope that this book will offer promising ideas and remarkable research contributions that will support continued advancements in cognition-based smart computing.

Editors' Biography

Sangita S. Chaudhari is currently working as a Professor and Head in the Department of Computer Science and Engineering at Ramrao Adik Institute of Technology, D. Y. Patil Deemed to be University, Navi Mumbai, India. She has published papers in international and national journals and conferences, as well as book chapters in the domains of information security, image processing, and pattern recognition. Her research interests include image processing, information security, geographical information systems, and remote sensing.

Pallavi Vijay Chavan is presently working as Professor in Department of Computer Engineering, Mukesh Patel School of Technology Management and Engineering, NMIMS University, Mumbai, India. She has 22 years of academic experience in data science and network security. Throughout her career, she has published research work in the areas of data science and security, contributing papers to journals and conferences of national and international repute.

Parikshit Mahalle is a Senior Member of IEEE and serves as Professor, Dean of Research and Development, and Head of the Department of Artificial Intelligence (AI) and Data Science at Vishwakarma Institute of Information Technology, Pune, India. He completed his Ph.D. at Aalborg University, Denmark, and subsequently worked as a postdoctoral researcher at CMI in Copenhagen, Denmark. He holds 12 patents and has published more than 200 research papers in national and international journals and conferences. His research interests include machine learning (ML), data science, algorithms, the Internet of Things, identity management, and security.

Pranav Mothabhau Pawar is currently working as an Assistant Professor in the Department of Computer Science at the Birla Institute of Technology and Science, Dubai. He developed a deep Q-learning–based asymptotically optimal solution to the spectrum collaboration problem and has published more than 40 papers in national and international journals and conferences.

Contributors

1. **Hammad Ansari**
   Ramrao Adik Institute of Technology
   Navi Mumbai, India

2. **Sumedha Bhagwat**
   Ramrao Adik Institute of Technology
   Navi Mumbai, India

3. **Ujwala Bharambe**
   Thadomal Shahani Engineering College
   Mumbai, India

4. **Smita Bharne**
   Ramrao Adik Institute of Technology
   Mumbai, India

5. **Om Bidve**
   Savitribai Phule Pune University
   Pune, Maharashtra

6. **Amar Buchade**
   BRACT's Vishwakarma Institute of Technology Kondhwa
   Pune, India

7. **Sangita S. Chaudhari**
   Ramrao Adik Institute of Technology
   Navi Mumbai, India

8. **Ashwini Dalvi**
   K. J. Somaiya College of Engineering
   Mumbai, India

9. **Radhika Gaikwad**
Savitribai Phule Pune University
Pune, Maharashtra

10. **Rajiv Iyer**
Amity University
Noida, India

11. **Pradip M. Jawandhiya**
P R Pote College of Engineering and Management
Amravati, India

12. **Pratham Joshi**
Ramrao Adik Institute of Technology
Mumbai, India

13. **Ahmed M. Khedr**
University of Sharjah
Sharjah, United Arab Emirates

14. **Vinit Kotak**
Shah & Anchor Kutchhi Engineering College
Mumbai, India

15. **Parikshit Mahalle**
BRACT's Vishwakarma Institute of Technology
Pune, India

16. **Manimala Mahato**
Shah & Anchor Kutchhi Engineering College
Mumbai, India

17. **Vedprakash Maralapalle**
L.S. Raheja School of Architecture
Mumbai, India

18. **Anirudh D. Maurya**
University of Mumbai
Mumbai, India

19. **Aatmaj Kaustubh Mhatre**
K. J. Somaiyya College of Engineering
Mumbai, India

20. **Ankan Mondal**
University of Mumbai
Mumbai, India

21. **Sankalp Naranje**
BRACT's Vishwakarma Institute of Technology
Pune, India

22. **Pragati Nehete**
Savitribai Phule Pune University
Pune, Maharashtra

23. **Leenata Parab**
Pillai College of Engineering
Mumbai, India

24. **Rachana Patil**
Pimpri Chinchwad College of Engineering
Pimpri-Chinchwad, India

25. **Kaushal Patil**
University of Southern California
Los Angeles, California

26. **Pranav M. Pawar**
Birla Institute of Technology and Science Pilani
Dubai, United Arab Emirates

27. **Pravija Raj Patinjare Veetil**
Birla Institute of Technology and Science Pilani
Dubai, United Arab Emirates

28. **Rekha Ramesh**
Kozhikode, Kerala

29. **Devesh Rathi**
BRACT's Vishwakarma Institute of Technology
Pune, India

30. **Kaustubh Rathod**
Ghatanji, Yavatma yl

31. **Rajani Sajjan**
MIT ADT University
Loni Kalbhor, India

32. **Irfan Siddhavatam**
K. J. Somaiyya College of Engineering
Mumbai, India

33. **Kahkashan Siddavatam**
University of Mumbai
Mumbai, India

34. **Radha Shirbhate**
Pankaj Laddhad Institute of Technology and Management Studies
Yelgaon, India

35. **Mansi Subhedar**
Pillai HOC College of Engineering and Technology
Maharashtra, India

36. **Nishant Wanjari**
Ramrao Adik Institute of Technology
Mumbai, India

Chapter 1

AI-Driven Optimization of Multi-Objective Task Scheduling in Fog-Assisted IoT Environments

Patinjare Veetil Pravija Raj, Pranav M. Pawar, and Ahmed M. Khedr

1.1 INTRODUCTION

The widespread adoption of the Internet of Things (IoT) in modern times has become essential for a multitude of industries and applications, spanning multiple domains and functionalities [1, 2]. While the need for IoT devices to have robust computing capabilities is evident, they are inherently constrained by factors such as limited battery life, processing, and storage ability. The broad adoption of IoT has resulted in a significant expansion in the scope of connected devices, which triggered network congestion and raised the demand for effective task offloading techniques [3, 4]. Consequently, there has been a growing trend towards task offloading from IoT to cloud servers as a means to mitigate these challenges. Cloud computing is a versatile infrastructure technology that allows users to utilize various computing resources and services on demand. Utilizing the internet, users can access these resources from anywhere, at any time. Within a cloud computing environment, tasks such as data processing and storage are handled by specialized cloud servers, allowing for efficient and flexible management of computing tasks and data. Task offloading represents the process of moving tasks from IoT to cloud for processing in order to circumvent the resource limitations faced by IoT nodes [5]. IoT tasks can be carried out more efficiently by making use of the cloud's robust processing and storage capabilities. However, despite these benefits, there are drawbacks to offloading all tasks to the cloud, including higher latency and congestion in the network due to the reliance on high bandwidth and cloud latency. Fog computing, which emerged as a novel paradigm aimed at bridging the gap between the IoT and cloud, facilitates lower latency and enhanced performance [6]. Offloading tasks to fog nodes within a cloud-fog environment has demonstrated its effectiveness in decreasing response times. With the capability of bringing computation resources closer to the network edge, fog computing minimizes the latency and alleviates network congestion, enabling a more efficient task offloading scheme for IoT.

DOI: 10.1201/9781003501220-1

As demands grow, researchers have devised innovative strategies to optimize data storage and processing in proximity to IoT devices, thereby enhancing the efficient utilization of network bandwidth and cloud resources [7, 8]. Consequently, the fog computing model emerges as a promising alternative that enhances task offloading efficiency in IoT environments. By assigning tasks to fog nodes, IoT devices effectively address latency constraints, resulting in significant reductions in power consumption. Task offloading yields numerous user-centric advantages, including decreased latency, enhanced reliability, and improved energy efficiency of IoT devices. The field of task offloading encompasses a diverse range of approaches, with recent years witnessing the emergence of various methods specifically designed to treat it as an optimization challenge [9, 10]. To address the challenges associated with task offloading, researchers have undertaken studies exploring different artificial intelligence (AI) optimization techniques. Designing effective task offloading criteria for fog computing presents a substantial challenge, particularly in efficiently utilizing fog resources to avoid fog nodes from becoming overloaded due to a large volume of offloaded tasks. Moreover, selecting the best fog node for offloading the tasks poses a significant challenge [11]. The existing task offloading models focus on reducing task delay and expended energy by prioritizing parameters including energy usage, node priority, and average energy of fog nodes. However, these models often face limitations in achieving a balance between the effectiveness of task offloading and energy optimization. Inadequate resource allocation or inefficient task offloading strategies can result in underutilization or overloading of fog nodes. Therefore, optimizing task allocation to ensure balanced workload distribution across fog nodes is crucial for maximizing system performance. Moreover, minimizing communication delays within the IoT-fog-cloud environment remains challenging due to varying network conditions and distances. Scalability challenges, stemming from limitations in fog node capacity, network congestion, and computational overhead, further complicate the task offloading process. Addressing these challenges necessitates the development of innovative approaches that carefully consider the unique characteristics of Fog-enabled IoT environments.

In this study, the task offloading problem for Fog-assisted IoT applications is examined, and an artificial intelligence (AI)-driven solution, termed the Multi-Objective Task-scheduling in Fog-assisted IoT based on Mountain Gazelle Optimizer (MTFMGO), is developed for optimizing multi-objective task scheduling in Fog-assisted applications of IoT by adapting the Mountain Gazelle Optimizer (MGO) algorithm. MGO is a novel meta-heuristic algorithm that takes its cues from the hierarchy and group dynamics seen in the natural territory of mountain gazelles [12]. In comparison to similar recently established meta-heuristic approaches, the MGO has proved its competitive performance. The proposed solution incorporates a novel approach by adapting MGO for optimizing the multi-objective task scheduling problem by taking into account a number of quality parameters. Extensive simulations indicate the excellence of the MTFMGO method over existing techniques under various metrics. The primary contributions of this work are:

- We propose MTFMGO, a multi-objective strategy for task offloading, aimed at tackling the challenge of scheduling tasks from IoT nodes to the fog layer. By effective management of task allocation across fog nodes, MTFMGO aims to accomplish effective load balancing and optimize system performance.
- We adapt the MGO algorithm to provide an effective task-offloading solution prioritizing the reduction of response time and cost. This is achieved by efficiently distributing tasks across fog nodes, taking into account the overall costs and computation time.
- We evaluate the efficacy of the MTFMGO scheme using comprehensive experimental analysis and conducting comparative analysis with existing algorithms.

1.2 RELATED WORKS

This section reviews the recent techniques for task scheduling in Fog-assisted IoT applications. Researchers have introduced different approaches to address the task scheduling challenge and achieve favorable solutions [1–3].

In [13], the authors proposed a dynamic computation offloading strategy in terms of Lyapunov optimization. The performance metric considered here is the execution cost, which accounts for both task failure and execution latency. Another approach suggested in [14] aims to reduce latency and energy utilization by employing a resource provisioning method based on a joint optimization scheme. In [15], an energy and fairness-aware method is introduced, focusing on energy consumption, the average energy of fog nodes, and priority. A distributed fog framework for task scheduling in smart production line systems is demonstrated in the research proposed by [16]. Integrated Particle Swarm Optimization (PSO) and Ant Colony Optimization (ACO) techniques are used to reduce delay and energy usage. For scheduling tasks in fog-based systems, [17] recommends an innovative optimization method based on the Bees algorithm with an objective of reducing the response delay and storage allotted to user service requests. In [18], a two-step fairness scheme to offload tasks in a fog computing environment is devised to mitigate latency and energy usage. The technique initially identifies the best fog node, followed by task offloading to the selected node based on a fairness metric. However, the method did not consider the communication delay. On the other hand, [19] proposes a two-stage strategy employing Grey Wolf Optimizer (GWO) to tackle the offloading issue, aiming to minimize energy expenditure and task execution time. A parallel meta-heuristics-based dynamic task offloading method utilizing Lyapunov optimization for enhanced energy efficiency and latency reduction is devised in [20]. The work in [21] presents a priority-based offloading mechanism using multi-level feedback queuing, where tasks are prioritized based on deadlines, ultimately reducing processing time and latency. In [4], the ACO algorithm is employed to reduce the average delay in task offloading processes. [5] designed an optimization scheme based on a combination of Long Short-Term Memory (LSTM) and GWO for task prediction and optimal offloading, enabling the analysis of task delay and associated energy utilization. The work in [22] employed two meta-heuristic methods (ACO and PSO) for developing scheduling strategies with the objective of reducing response time for the tasks. In [23], authors introduced a PSO-based hybrid heuristic strategy to tackle task offloading challenges. This approach employs a three-stage heuristic technique, starting with maximizing cloud utilization, followed by the utilization of edge resources for faster response times, and finally optimizing task execution performance through selective re-offloading. Another efficient offloading strategy, proposed in [24], combines ACO and PSO algorithms to ensure load balance among fog nodes and minimize response times. Results indicate that ACO outperforms PSO in task offloading effectiveness and efficiency. In [6], a multi-objective PSO is employed to address offloading challenges, focusing on minimizing task delays using the branch and bound approach. For reducing energy loss and time consumption, [7] presents an alternative task offloading approach based on quantum behavior PSO to minimize delay and cost. In [8], an accelerated PSO method is proposed to handle task offloading as a multi-objective challenge, considering execution time and resource utilization while excluding cost considerations. Addressing task offloading challenges with an Artificial Bee Colony optimization-based (ABC-based) method, [9] aims to reduce response times. However, this approach focuses only on response time optimization without considering multi-objective optimization. Similarly, [11] proposes a smart ACO technique for optimizing the offloading time in IoT-fog applications, however, it does not incorporate multi-objective optimization considerations. Utilizing binary hybrid GWO, [25] addresses task offloading to minimize response rates. In contrast, [26] proposes an improved PSO-based model for task offloading and caching decisions to optimize offloading and cache decisions and reduce total costs.

In recent years, meta-heuristic algorithms have demonstrated success in addressing a variety of problems [1, 10, 27]. Additionally, various other optimization techniques, both in single-objective and multi-objective conditions, have been utilized to tackle diverse challenges [2, 28, 29]. As evident from the literature review, most of the existing scheduling methods focus on task offloading as a single-objective optimization strategy or fail to effectively address different key factors, such as response time, resource utilization, and cost. However, optimizing multiple Quality of Service (QoS) metrics is essential for achieving greater efficiency. In this paper, we introduce a novel multi-objective task-offloading approach by adapting the MGO algorithm. Recently, the MGO algorithm has gained allure among researchers for its efficacy. In contrast to other recently established meta-heuristic methods, MGO has shown competitive performance. Our proposed approach adapts the capabilities of MGO to effectively allocate tasks across fog nodes, considering the comprehensive costs and computation time for scheduling tasks in Fog-assisted IoT settings.

1.3 PROPOSED AI-DRIVEN MULTI-OBJECTIVE TASK SCHEDULING FRAMEWORK

This section provides the details of the proposed MTFMGO approach for effective scheduling of tasks in Fog-assisted IoT. We begin with the system model and problem description, followed by MGO background. Then, a detailed description of the MTFMGO scheme is presented.

1.3.1 System Design and Problem Description

The system model for task scheduling is shown in Figure 1.1. It consists of IoT devices with limited capabilities at the bottom layer, fog in the middle layer, and cloud at the top layer. Higher levels receive requests from IoT devices and process and store them there. The computing nodes within the fog layer manage task scheduling by applying scheduling algorithms to assess node resources and task attributes. Tasks requiring powerful processing and large amounts of data storage over long periods of time are designated for the cloud level. After task requests are processed, the results are sent to users or IoT devices so they can make data-driven decisions.

Assume that the fog layer has N fog nodes $\zeta_1, \zeta_2, \ldots, \zeta_N$, and C nodes in the cloud layer. The IoT layer contains M devices $D_1, D_2, D_3, \ldots, D_M$. Let d denote the total number of tasks with respect to the IoT devices, where:

$$\Gamma = \Gamma 1 + \Gamma 2 + \Gamma 3 + \ldots + \Gamma d \tag{1}$$

The tasks generated from IoT nodes can be completed either locally or offloaded to be processed by fog nodes within the fog layer. Given that there is no means to split tasks, each task can only be assigned to one fog node ζ_i. However, a single node ζ_i possesses the ability to manage multiple allocated tasks concurrently.

1.3.2 MGO Background

The MGO algorithm presents a novel optimization strategy and as stated earlier in the chapter, the approach is inspired by the social behaviors of

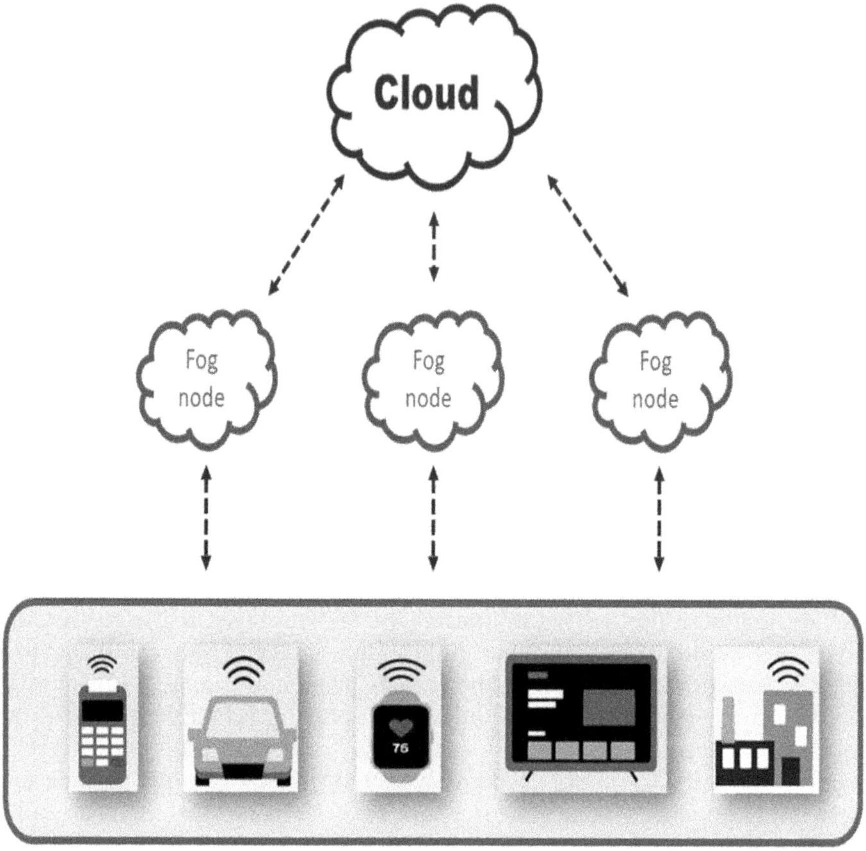

Figure 1.1 Internet of Things-Fog-Cloud System Model.

mountain gazelles (*mGz*) [12]. It is based on the collective social interaction principles seen in *mGz* and represents four major aspects of their life: (1) solitary territorial males; (2) maternity herds; (3) bachelor male herds; and (4) migration for food inquiry. Every iteration eliminates feeble gazelles and selects strong ones that produce high-quality results.

1. **Territorial solitary males** (mGzT): The equation that follows is used to model this behavior. The position vector of the best global result is denoted as *malegz* , and the values of *r*1 and *r*2 are randomly assigned to either 1 or 2.

$$\mathrm{mGzT} = \mathrm{malegz} - \left|(\mathrm{r1} \times \mathrm{BmH} - \mathrm{r2} \times \mathrm{X}(\mathrm{t})) \times \mathrm{F}\right| \times \mathrm{Cr} \tag{2}$$

The following formulas are used to calculate the coefficients BmH, F, and Cr:

$$\mathrm{BmH} = \mathrm{XrA} \times [\mathrm{rnd1}] + \mathrm{Mc} \times [\mathrm{rnd2}], \mathrm{rA} = [\mathrm{N/3}] \ldots \mathrm{N} \tag{3}$$

$$\mathrm{F} = \mathrm{N1}(\mathrm{D}) \times \exp\left(2 - \mathrm{itr} \times (2/\mathrm{Mitr})\right) \tag{4}$$

$$Cr = \begin{cases} (a+1) + rnd3 \\ \left(a \times N2(D)\right) \\ \left(rnd4(D)\right) \\ \left(N3(D) \times N4(D)^2 \times cos\left((rnd4 \times 2) \times N3(D)\right)\right. \end{cases} \tag{5}$$

$$a = -1 + itr \times \left(\frac{-1}{Mitr}\right) \tag{6}$$

where, XrA denotes a randomly chosen outcome within the rA interval, where N stands for the total count of gazelles and Mc indicates the average count of gazelles. The variables $rndi$ (i = 1, ..., 4) represent random values ranging between 0 and 1. The random variable $N1$ is drawn from a standard distribution, while random values $N2$, $N3$, and $N4$ are drawn from a normal distribution. The terms itr and $Mitr$ indicate the current and maximum iterations, respectively.

2. **Maternity herds** (*MM*)**:** The equation that follows is used to model this behavior.

$$MM = (BH + Cr) + (rz3 \times malegz - rz4 \times Xrgz) \times Cr \tag{7}$$

where, BH is calculated using Equation 3, Cr is obtained using Equation 5, $rz3$ and $rz4$ are random integers 1 or 2, $malegz$ is the global result location vector and $Xrgz$ is a random gazelle location vector, respectively.

3. **Bachelor male herds** (*MB*)**:** The equation that follows is used to model the MB behavior, where $malegz$ is the best male gazelle position, BH is determined using Equation 3, $rnd6 \in (0, 1)$ denotes a random number, Cr is calculated in accordance with Equation 5, and $X(t)$ is the current gazelle position. Dn is computed using Equation 9, $rz5$ and $rz6$ denotes a random value 1 or 2, respectively.

$$MB = \left(X(t) - Dn\right) + (rz5 \times malegz - rz6 \times BH) \times Cr \tag{8}$$

$$Dn = \left(\left|X(t)\right| + \text{l}malegz\right) \times (2 \times rnd6 - 1) \tag{9}$$

4. **Migration for food search** (MF): The equation that follows is used to model the MF behavior.

$$MF = (\text{ul} - \text{ll}) \times \text{rnd7} + \text{lbd} \tag{10}$$

where the problem's upper and lower bounds are indicated by ul and ll, respectively, and $rnd7$ is a randomly selected number between 0 and 1.

1.3.3 Proposed MTFMGO Scheme

This section provides the details of the proposed multi-objective methodology for offloading IoT tasks to fog nodes, by adapting the MGO algorithm. The IoT tier comprises a set of distributed devices, D_i, $i = 1 \ldots M$ where each device D_i produces data at rate (κ_i). Consider Γi denotes a task generated by an IoT device with a size Γszi. Let ηf denote the fog bandwidth, and ηc denote the cloud bandwidth.

1.3.3.1 Objective Function Formulation

As a multi-objective optimization problem, the task offloading issue can be expressed as follows:

- Based on Equation 11, the communication time between the fog node ζ_j and the IoT device D_i is calculated. The communication time between fog and cloud (Tfc) is estimated using the Equation 12 by considering the cloud latency $\psi_{j,c}$ and time taken for data transfer from fog to cloud, respectively, as follows:

$$Tdf(i,j) = \frac{\Gamma szi}{\eta f} + \psi_{i,j} \tag{11}$$

where $\psi_{i,j}$ represents the latency between D_i and ζ_j respectively.

$$Tfc(i,j) = \frac{\Gamma szi}{\eta c} + \psi_{j,c} \tag{12}$$

Therefore, the total communication time can be derived as below:

$$Tcom = Tdf(i,j) + Tfc(i,j) \tag{13}$$

- The **service time** ($\Upsilon_{i,j}$) of the task Γi offloaded to ζ_j is computed using Equation 14 and the service time ($\Upsilon_{j,c}$) from fog to cloud is estimated following the Equation 15, respectively, as follows:

$$\Upsilon_{i,j} = \frac{1}{\delta_{i,j} - \kappa_i} \quad (14)$$

$$\Upsilon_{j,c} = \frac{1}{\delta_{j,c} - \kappa_{j,c}} \quad (15)$$

Therefore, the total service time can be derived as below:

$$Tser = \Upsilon_{i,j} + \Upsilon_{j,c} \quad (16)$$

where $\delta_{i,j}$, $\delta_{j,c}$ *and* κ_i, $\kappa_{j,c}$ represent the corresponding service rates and task arrival rates, respectively.

- The **overall response time** ($TRes$) for the tasks triggered by the IoT device is estimated based on the total communication time and the service time as specified in Equation 17.

$$TRes = Tcom + Tser \quad (17)$$

- **Processing cost computation:** The time for processing a task is computed in terms of its size Γsz_i and the capacity (ϵ_j) of the fog node, as below.

$$Tproc = \frac{\Gamma szi}{\epsilon_j} \quad (18)$$

Therefore, the processing cost ($Cproc$) for task Γ is computed using Equation 19,

$$Cproc = \frac{Tproc \times C\zeta i}{Tp} \quad (19)$$

where $C\zeta i$ is the processing cost of the fog node ζi for a time interval Tp. Memory cost ($Cmem$) of the task Γ is determined as follows:

$$Cmem = \frac{Tproc \times SC\zeta i}{Tq} \tag{20}$$

where, $SC\zeta i$ is the storage cost of the fog node ζ_i for time interval Tq. Based on the above, the total processing cost ($T\ otCproc$) can be derived as follows:

$$T\ otCproc = Cproc + Cmem \tag{21}$$

- The **latency cost** ($Clat$) is calculated as follows, which depends on the communication cost $Ccom$ for the interval Tr

$$Clat = \frac{\Lambda \times Ccom}{Tr} \tag{22}$$

where Λ is given by,

$$\Lambda = Cproc + \Gamma szi \tag{23}$$

- Therefore, the total cost for offloading the task $TCoff$ is determined as follows.

$$TCoff = TotCproc + Clat \tag{24}$$

The objective function is defined in Equation 25, which aims to maximize response time and task planning costs.

$$Minimize \sum_{\Gamma i=1}^{d} \alpha \cdot TRes + \beta \cdot TCoff \tag{25}$$

where α *and* β are constant weighting parameters corresponding to the response time and the task's offloading cost.

1.3.3.2 Algorithm Description

To provide an efficient task-offloading solution, the proposed MTFMGO scheme adapts and employs the MGO. In the proposed scheme, the initial population generation of MGO involves the utilization of chaotic mapping, replacing the conventional random generation typically used in actual MGO. The application of chaotic sequence prevents premature convergence while simultaneously enhancing the initial population within the actual MGO method. The detailed description of MTFMGO is as follows:

Algorithm 1.1 MTFMGO Algorithm

1. **Input:** Tasks list: Γ=Γ_1, Γ_2, Γ_3, . . ., Γ_d, Fog nodes N : $\zeta 1$, $\zeta 2$, ..., ζN , C cloud nodes, MGO parameters: population size MN, maximum iterations $Mitr$.
2. **Output:** Best offloading result.
3. **BEGIN**
4. Initialize the mountain gazelle population using the proposed approach (Equation 26).
5. **while** $Mitr$ not reached **do**
6. Evaluate each gazelle based on fitness following the objective function (Equation 25).
7. Identify the best and worst gazelle.
8. **for** each gazelle in the population **do**
9. Determine $mGzT$, MM, MB, MF (Equations 2, 7, 8 and 10).
10. Compute fitness scores and add them to the population.
11. **end for**
12. Sort the population based on fitness.
13. Retain strong N gazelles with high-quality results.
14. Revise the best and worst gazelle.
15. Increment the iteration count.
16. **end while**
17. Return the best gazelle found.
18. **END**

The pseudocode of MTFMGO is given in Algorithm 1.1. It accepts the input: the tasks list: $\Gamma = \Gamma_1, \Gamma_2, \Gamma_3, \ldots, \Gamma_d$, Fog nodes N: $\zeta_1, \zeta_2, \ldots, \zeta_N$, C cloud nodes, and the MGO parameters. The gazelle population is initialized following the proposed initialization approach using piecewise chaotic mapping given in Equation 26, where $\rho = 0.4$ is the control parameter [Steps 1–4].

$$X(t+1) = \begin{cases} \dfrac{X(t)}{\rho}, & 0 \le X(t) < \rho \\ \dfrac{X(t)-\rho}{0.5-\rho}, & \rho \le X(t) < 0.5 \\ \dfrac{1-\rho-X(t)}{0.5-\rho}, & 0.5 \le X(t) < 1-\rho \\ \dfrac{1-X(t)}{\rho}, & 1-\rho \le X(t) < 1 \end{cases} \tag{26}$$

Each gazelle's location is examined and adjusted based on fitness [Step 6]. The best and worst individuals are then identified [Step 7]. Determine *mGzT*, *MM*, *MB*, and *MF* for each individual using the Equations 2, 7, 8, and 10, respectively [Steps 9–14]. After assessing their fitness, include them in the population [Steps 8–10]. Only the strong *N* gazelles that offer superior solutions will be retained after the population has been sorted based on fitness. Revise the gazelles ranked worst and best [Steps 11–14]. Increase the iteration count until *Mitr* is attained [Step 15–16]. Ultimately, the output is the optimal gazelle corresponding to the best offloading result [Step 17–18].

1.4 SIMULATION RESULTS

The simulation results for the MTFMGO method are discussed in this section. Experiments are performed using MATLAB. The MTFMGO method is compared with the existing approaches: the ACO-based method [22], the ABC-based method [9], as well as the standard MGO [12] and the standard Sparrow Search Algorithm (SSA) [30] algorithms. Table 1.1 lists the various simulation parameters that are used. The results are discussed as follows:

- **Average response time** ($AvgT_{res}$): The average time of response for the different compared algorithms with varying number of tasks is shown in Figure 1.2. The results reveal that MTFMGO reduces the $AvgT_{res}$ as the tasks count increases. The standard MGO method shows higher response times indicating lower performance compared to MTFMGO. SSA and ABC have relatively higher response times as the number of tasks increases, and ACO has the least performance. It is evident that the MTFMGO minimizes the $AvgT_{res}$ by 4%, 9%, 11%, and 15% compared to MGO, SSA, ACO, and ABC methods, when the number of tasks is 1250.

Table 1.1 Parameter Settings

Parameter	*Value*
Tasks count	500–2000
Fog nodes	20
Task size	250–1000 KB
Search agents count	50
Max.iterations	100
η_c	12 MB/s
Υ	50–200
κ	10–50
$\psi_{i,j}$	20 ms
$\psi_{j,c}$	30 ms

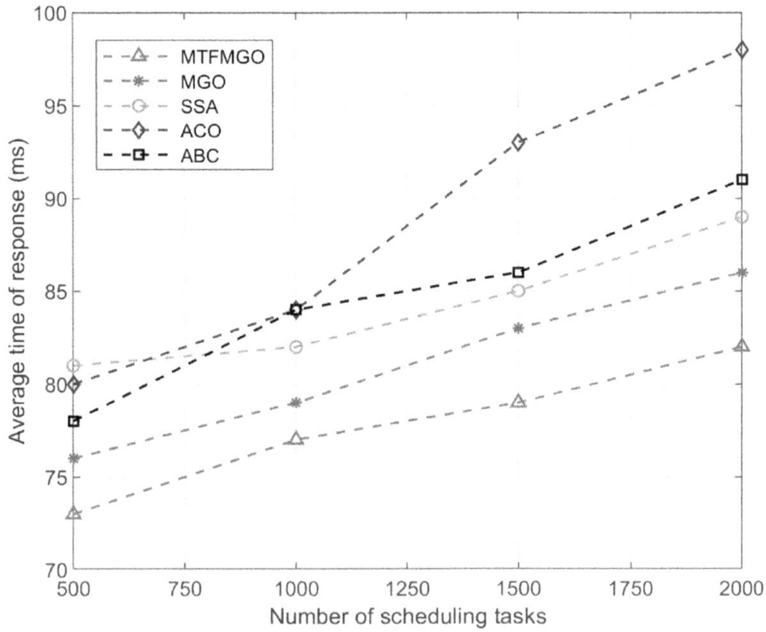

Figure 1.2 Comparison of Average Response Time ($AvgT_{res}$) Results.

The chaotic map initialization approach enhances the initial population and the overall diversity of the population, contributing to MTFMGO's superior performance. The improvements in MTFMGO, including modifications to the standard MGO and the development of a more effective objective function, result in its better performance over other algorithms.

- **Average task offloading cost** ($AvgC_{off}$)**:** Figure 1.3 shows the average computational cost of offloading tasks to fog nodes. It is obvious that costs tend to rise as the number of tasks surges. This demonstrates the direct relationship between the number of tasks and the offloading cost. According to the findings, MTFMGO has the lowest offloading costs and the slowest rate of cost growth. In comparison to MGO, SSA, ACO, and ABC approaches, MTFMGO minimizes the cost by an average of 7%, 11%, 16%, and 17%.
    - The **standard deviation of the response time** (Std_{Res})**:** Standard deviation represents the distribution of tasks among the fog nodes. The proposed MTFMGO shows enhanced performance in the Std_{Res} result and better stability as illustrated in Figure 1.4. This indicates that MTFMGO distributes the tasks efficiently

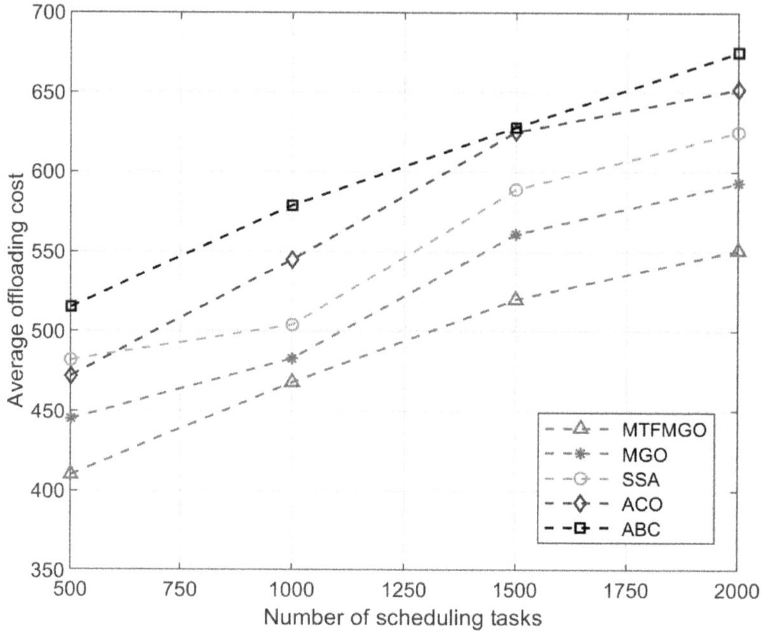

Figure 1.3 Comparison of Average Task Offloading Cost ($AvgC_{off}$) Results.

among the Fog nodes. The Std_{Res} of MTFMGO is fairly lower than MGO, SSA, ACO, and ABC methods.

- **Degree of imbalance** (Deg_{im})**:** The MTFMGO algorithm achieved the least degree of imbalance between offloaded tasks when compared to MGO, SSA, ACO, and ABC methods. The results are demonstrated in Figure 1.5. MTFMGO exhibits superior performance in minimizing the average degree of imbalance in offloaded tasks compared to MGO, which shows higher imbalance values. ABC, on the other hand, demonstrates the highest imbalance results among all methods considered. This highlights the effectiveness of MTFMGO in optimizing task scheduling by significantly reducing imbalance and outperforming the other methods.

1.5 CONCLUSION

This study considers the task offloading problem in Fog-assisted IoT environments as a multi-objective optimization challenge. By adapting the MGO algorithm, this research proposed a novel approach to task offloading optimization, designated MTFMGO, to reduce offloading costs and reaction times. Simulation results demonstrate that MTFMGO exhibits superior

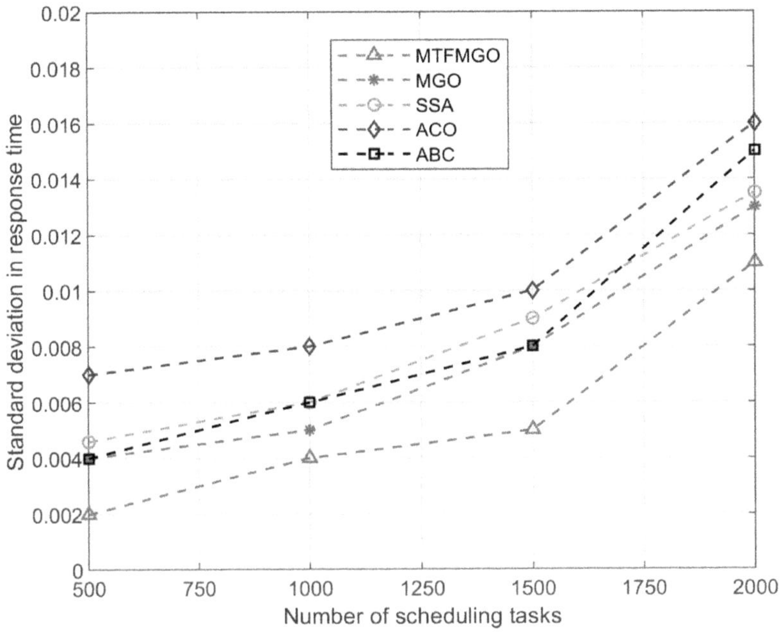

Figure 1.4 Comparison of Standard Deviation of the Response Time (Std_{Res}) Results.

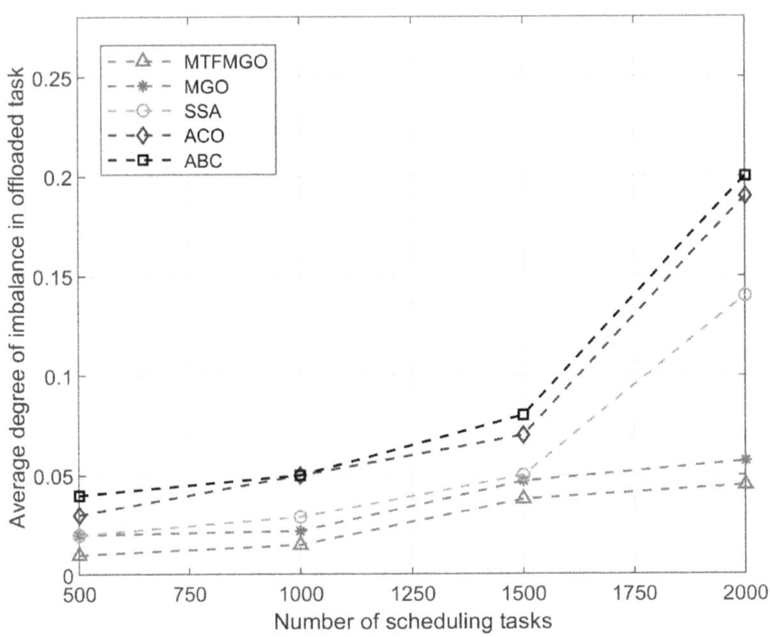

Figure 1.5 Comparison of Degree of Imbalance (Deg_{im}) Results.

performance in addressing the offloading challenge and effectively distributes tasks among fog nodes, leading to reduced costs and minimized response times. Compared to traditional methods such as MGO, SSA, ACO, and ABC methods, the proposed approach offers a more effective solution for optimizing task offloading in IoT-based systems.

REFERENCES

[1] Liu, Q., Kosarirad, H., Meisami, S. Alnowibet, K. A., & Hoshyar, A. N. (2023). An optimal scheduling method in IoT-fog-cloud network using combination of aquila optimizer and African vultures' optimization, *Processes*, 11(4), 1162.

[2] Dai, X., Xiao, Z., Jiang, H., Alazab, M., Lui, J. C., Min, G., Dustdar, S., & Liu, J. (2022). Task offloading for cloud-assisted fog computing with dynamic service caching in enterprise management systems, *IEEE Transactions on Industrial Informatics*, 19(1), 662–672.

[3] Wang, Y., Han, X., & Jin, S. (2023). Map based modeling method and performance study of a task offloading scheme with time-correlated traffic and vm repair in mec systems, *Wireless Networks*, 29(1), 47–68.

[4] Song, S., Ma, S., Yang, L., Zhao, J., Yang, F., & Zhai, L. (2022). Delay-sensitive tasks off loading in multi-access edge computing, *Expert Systems with Applications*, 198, 116730.

[5] Sharma D. K., & Singh, N. P. (2024). Task prediction and optimal offloading decisions for fog computing cloud network, *5G and Beyond Wireless Networks*, 207–220, CRC Press.

[6] Song, S., Ma, S., Zhu, X., Li, Y., Yang, F., & Zhai, L. (2023). Joint band width allocation and task offloading in multi-access edge computing, *Expert Systems with Applications*, 217, 119563.

[7] Chen, L., Zhang, D.-G., Zhang, J., Zhang, T., Wang, W.-J., & Cao, Y.-H. (2023). A novel offloading approach of IoT user perception task based on quantum behavior particle swarm optimization, *Future Generation Computer Systems*, 141, 577–594.

[8] Alfakih, T., Hassan, M. M., & Al-Razgan, M. (2021). Multi-objective accelerated particle swarm optimization with dynamic programing technique for resource allocation in mobile edge computing, *IEEE Access*, 9, 167503–167520.

[9] Babar, M., Khan, M. S., Din, A., Ali, F., Habib, U., & Kwak, K. S. (2021) Intelligent computation offloading for IoT applications in scalable edge computing using artificial bee colony optimization, *Complexity*, 2021(1), 5563531.

[10] Qiao, L., Naderi, S., Ahmadi, M., & Mirjalili, S. (2022). A workflow scheduling in cloud environment using a combination of moth-flame and SALP swarm algorithms, *SSRN Electronic Journal*. 10.2139/ssrn.4216421.

[11] Kishor A. & Chakarbarty, C. (2021). Task offloading in fog computing for using smart ant colony optimization, *Wireless Personal Communications*, 127(2), 1683-1704.

[12] Abdollahzadeh, B., Gharehchopogh, F. S., Khodadadi, N., & Mirjalili, S. (2022). Mountain gazelle optimizer: A new nature-inspired metaheuristic algorithm for global optimization problems, *Advances in Engineering Software*, 174, 103282.

[13] Mao, Y., Zhang, J., & Letaief, K. B. (2016). Dynamic computation offloading for mobile-edge computing with energy harvesting devices, *IEEE Journal on Selected Areas in Communications*, 34(2), 3590–3605.

[14] Sardellitti, S., Scutari, G., & Barbarossa, S. (2015). Joint optimization of radio and computational resources for multicell mobile-edge computing, *IEEE Transactions on Signal and Information Processing over Networks*, 1(2), 89–103.

[15] Zhang, G., Shen, F., Liu, Z., Yang, Y., Wang, K., & Zhou, M. (2018). FEMTO: Fair and energy-minimized task offloading for fog-enabled IoT networks, *IEEE Internet of Things Journal*, 6(3), 4388–4400.

[16] Wang J. & Li, D. (2019). Task scheduling based on a hybrid heuristic algorithm for smart production line with fog computing, *Sensors*, 19(5), 1023.

[17] Bitam, S., Zeadally, S., & Mellouk, A. (2018). Fog computing job scheduling optimization based on bees swarm, *Enterprise Information Systems*, 12(4), 373–397.

[18] Zhang, G., Shen, F., Yang, Y., Qian, H., & Yao, W. (2018) *Fair task offloading among fog nodes in fog computing networks*, 1–6.

[19] Hashemi, S. M., Sahafi, A., Rahmani, A. M., & Bohlouli M. (2022). GWO-SA: Gray wolf optimization algorithm for service activation management in fog computing, *IEEE Access*, 10, 107846–107863.

[20] AlShathri, S. I., Chelloug, S. A., & Hassan, D. S. M. (2022). Parallel meta-heuristics for solving dynamic offloading in fog computing, *Mathematics*, 10(8).

[21] Adhikari, M., Mukherjee, M. & Srirama, S. N. (2019). DPTO: A deadline and priority-aware task offloading in fog computing framework leveraging multilevel feedback queueing, *IEEE Internet of Things Journal*, 7(7), 5773–5782.

[22] Hussein, M. K. & Mousa, M. H. (2020). Efficient task offloading for IoT-based applications in fog computing using ant colony optimization, *IEEE Access*, 8, 37191–37201.

[23] Gao, T., Tang, Q., Li, J., Zhang, Y., Li, Y., & Zhang, J. (2022). A particle swarm optimization with levy´ flight for service caching and task offloading in edge-cloud computing, *IEEE Access*, 10, 76636–76647.

[24] Canaliand, C. & Lancellotti, R. (2019). GASP: Genetic algorithms for service placement in fog computing systems, *Algorithms*, 12(10), 201.

[25] Liu M. & Zhang, J. (2022). *Research on task offloading strategy for ultra dense edge computing network*, 587–592.

[26] Yuan K. & Chen, S. (2022). *Improved particle swarm optimization-based computation offloading and caching decision for internet of things*, 1068–1073.

[27] Salehnia T. & Fathi, A. (2021). Fault tolerance in LWT-SVD based image watermarking systems using three module redundancy technique, *Expert Systems with Applications*, 179, 115058.

[28] Xu, X., Lin, Z., Li, X., Shang, C., & Shen, Q. (2022). Multi-objective robust optimisation model for MDVRPLS in refined oil distribution, *International Journal of Production Research*, 60(22), 6772–6792.

[29] Li, B., Tan, Y., Wu, A.-G., & Duan, G.-R. (2021). A distributionally robust optimization based method for stochastic model predictive control, *IEEE Transactions on Automatic Control*, 67(11), 5762–5776.

[30] Awadallah, M. A., Al-Betar, M. A., Doush, I. A., Makhadmeh, S. N., & Al-Naymat, G. (2023). Recent versions and applications of sparrow search algorithm, *Archives of Computational Methods in Engineering*, 30(5), 2831–2858.

Chapter 2

Transforming Healthcare through Cognitive Computing and Smart Computing

A Comprehensive Overview

Rajiv Iyer, Vedprakash Maralapalle, and Leenata Parab

2.1 INTRODUCTION

Technological advancements have made cognitive and smart computing transformative forces in various industries, especially healthcare. This convergence profoundly impacts patient care, clinical decision-making, and operational efficiency. This introduction explores the fundamental concepts, challenges, and reasons for integrating intelligent technologies into healthcare. By examining the applications, frameworks, and implications of cognitive and smart computing, we aim to reveal their potential to improve patient outcomes, enhance provider experiences, and initiate a new era of healthcare excellence. In recent years, computing has shifted, enabling machines to demonstrate human-like intelligence. This evolution has led to two key concepts: cognitive computing and smart computing. Cognitive computing is a branch of artificial intelligence (AI) that mimics human thought processes. It uses machine learning (ML), natural language processing (NLP), computer vision, and pattern recognition to understand and learn from data. Unlike traditional computing, cognitive systems adapt based on user interactions and feedback. They emphasize context awareness and can interpret complex information to make real-time decisions. Examples include virtual assistants, recommendation systems, and medical diagnostic tools. Smart computing, also known as pervasive or ubiquitous computing, creates intelligent environments that sense, analyze, and respond to user needs. It focuses on interconnectedness across devices, sensors, and networks, epitomized by the Internet of Things (IoT). Smart computing integrates IoT, cloud computing, edge computing, and data analytics to enhance convenience and efficiency. It orchestrates devices and services to provide personalized experiences and optimize resources, seen in smart homes, cities, and transportation systems.

Cognitive and smart computing are transforming modern computing. Cognitive computing imbues systems with human-like abilities, while smart computing creates interconnected ecosystems to enhance daily life.

DOI: 10.1201/9781003501220-2

Understanding these technologies allows us to innovate and amplify human potential.

2.2 OVERVIEW OF THE HEALTHCARE LANDSCAPE AND CHALLENGES

The healthcare sector is undergoing transformation through smart and cognitive computing. Understanding the current landscape and challenges is crucial.

1. **Current healthcare landscape:** The healthcare industry involves various stakeholders like providers, patients, payers, and regulators. Traditionally, it relied on reactive interventions and fragmented data systems. Digital technologies now offer opportunities for proactive, personalized care.
2. **Challenges in healthcare:** Despite advancements, the sector faces challenges in efficiency, affordability, and accessibility, including:
   - *Data silos and interoperability:* Data is often siloed, hindering seamless information exchange and holistic patient care. Achieving interoperability is essential for leveraging smart and cognitive computing.
   - *Quality and safety:* Quality and safety are the bedrocks of effective healthcare. Smart computing technologies have been rising as strong helpers for enhancing clinical decision-making, reducing errors and ultimately bettering patient results. For instance, electronic health records (EHRs) together with clinical decision support systems (CDSS) when integrated can go a long way in enabling instantaneous access to patients' data, proposing proper medication and treatment plans as well as making alarms for any possible medication allergies or drug interactions [1].
   - *Rising costs:* Escalating costs strain healthcare systems globally. Smart and cognitive computing can optimize resource allocation and improve outcomes.
3. **Regulatory compliance and privacy:** Navigating regulatory frameworks while leveraging emerging technologies requires balancing innovation and compliance. Smart and cognitive computing offer innovative solutions to enhance efficiency and patient-centric care.

2.2.1 Rationale for the Integration of Cognitive and Smart Computing in Healthcare

The integration of cognitive and smart computing promises to revolutionize healthcare delivery and improve patient outcomes. Cognitive computing leverages extensive datasets, including genomic information, to enable precision medicine that is customized to meet individual patient needs [2]. ML algorithms enable proactive healthcare by anticipating adverse events and enabling timely interventions [3]. Cognitive computing synthesizes information to provide actionable insights, enhancing diagnostic accuracy and treatment efficacy. Smart computing technologies, like mobile health applications and wearable devices, empower patients to actively participate in their healthcare [4].

Cognitive and smart computing shift healthcare from reactive to proactive, personalized, and patient-centric care.

2.2.2 Overview of Cognitive Computing Technologies

Cognitive computing mimics human intelligence using technologies like NLP, ML, knowledge representation, and advanced analytics. These technologies help machines understand unstructured data and learn from experience. Table 2.1 demonstrates the use of innovations in healthcare technology.

1. **Natural language processing (NLP):** NLP algorithms interpret human language, aiding in parsing clinical notes and medical literature. For example, NLP can extract information from EHRs to assist in patient care [5].
2. **Machine learning (ML):** ML algorithms identify patterns and make predictions from data. In healthcare, they analyze patient demographics and treatment outcomes to provide insights. Predictive models can forecast readmissions and optimize treatment protocols.

2.2.3 Case Studies and Examples

Case studies and examples provide tangible evidence of cognitive computing's impact on healthcare delivery, showcasing successful implementations and outcomes.

Examples:

1. **IBM Watson Health:** Watson for Oncology assists oncologists in diagnosing and treating cancer. A study in the *Journal of Clinical Oncology* found that Watson's treatment plans agreed with expert oncologists' decisions in 96% of breast cancer cases.
2. **Google's DeepMind Health:** DeepMind's Streams app uses AI to analyze patient data and prioritize alerts for deteriorating patients. At the Royal Free London NHS Foundation Trust, Streams reduced the time to detect acute kidney injury (AKI) from hours to minutes, enabling prompt clinician intervention and improved patient outcomes.
3. **Memorial Sloan Kettering Cancer Centre's Oncology (MSKCC's) Expert Advisor:** MSKCC's Oncology Expert Advisor provides

Table 2.1 Innovations in Healthcare Technology

Case Study	*Problem Addressed*	*Technology Utilized*	*Outcomes Achieved*
IBM Watson for Oncology	Diagnosing and treating cancer	Cognitive computing system (IBM Watson)	Recommended treatment plans concurred with expert oncologists' decisions in 96% of breast cancer cases.
Google's DeepMind Health–Streams	Prioritizing alerts for deteriorating patients	AI algorithms	Reduced time to detect acute kidney injury (AKI) from hours to minutes, enabling prompt interventions and improving patient outcomes.
Memorial Sloan Kettering Cancer Center's Watson for Genomics	Providing personalized treatment recommendations for cancer patients	Cognitive computing system (IBM Watson)	Identified potentially actionable alterations in 99% of patient cases, demonstrating efficacy in guiding precision oncology treatments.
Mayo Clinic's Clinical Decision Support System	Improving diagnostic accuracy and treatment outcomes	Cognitive computing-based clinical decision support system	Reduced diagnostic errors by 30%, leading to shorter hospital stays and lower mortality rates.
University of Pittsburgh Medical Center's Prescriptive Analytics Platform	Optimizing healthcare operations and resource allocation	Prescriptive analytics platform powered by cognitive computing	Reduced emergency department wait times by 25% and increased patient satisfaction scores by 15% (UPMC, internal data).

personalized cancer treatment recommendations by analyzing medical records, genetic profiles, and treatment histories. A study in the *Journal of Clinical* Oncology found improved adherence to guidelines and better patient outcomes [6].

4. **University of California, San Francisco's (UCSF's) CareWeb:** UCSF's CareWeb integrates patient data from EHRs, genomic databases, and clinical trials to support oncology decision-making. NLP algorithms extract relevant information from clinical notes and research articles, providing oncologists with comprehensive patient information. A study in the *Journal of the American Medical Informatics Association* showed that CareWeb led to more informed treatment decisions, reduced delays, and improved patient outcomes [7].

Here is a comparative overview of transformative technologies utilized in oncology diagnosis, treatment, and clinical decision-making processes. Smart computing is revolutionizing healthcare with data-driven decision-making, personalized care, and improved patient outcomes. This section explores its principles, applications, and impact.

2.3 OVERVIEW OF SMART COMPUTING TECHNOLOGIES

Smart computing in healthcare involves advanced data processing and AI to enhance care, streamline tasks, and optimize resources. Key technologies include:

1. **Electronic health records (EHRs):** EHRs digitize patient health information, facilitating secure access and data-driven decision-making while improving care coordination.
2. **Wearable devices:** Wearables like fitness trackers and smartwatches monitor health metrics, promote preventive care, and enable remote patient monitoring [8].
3. **Telemedicine:** Telemedicine uses telecommunications for remote healthcare services, improving access, especially in rural or underserved areas [9].
4. **Artificial intelligence (AI):** AI analyzes medical data to identify patterns, make predictions, and assist in clinical decisions. It includes applications like image recognition and predictive analytics.
5. **Internet of Medical Things (IoMT):** IoMT consists of interconnected medical devices and sensors that enable continuous monitoring, remote diagnostics, and personalized interventions [10].

2.3.1 Applications of Smart Computing in Healthcare

Smart computing revolutionizes patient care, clinical decision-making, and healthcare management through various applications:

1. **Clinical decision support systems (CDSS):** CDSS use AI to provide evidence-based recommendations and alerts, aiding in disease diagnosis, treatment selection, and error prevention [11].
2. **Remote patient monitoring:** Wearable devices and IoMT sensors track health metrics outside traditional settings, enabling early detection, timely interventions, and reduced hospital readmissions.
3. **Personalized medicine:** Smart computing integrates genetic, clinical, and lifestyle data to tailor treatments, improving outcomes and minimizing adverse effects.
4. **Health informatics:** Health informatics platforms analyze EHR data to generate insights into population health, disease prevalence, and treatment outcomes, informing public health and policy.

2.3.2 Case Studies

Case studies demonstrate smart computing's impact on healthcare delivery, showcasing successful implementations and outcomes.

Case Study 1: Implementation of Clinical Decision Support System [12]

In a large tertiary care hospital, a CDSS was implemented to reduce medication errors and enhance adherence to evidence-based guidelines. Integrated with the hospital's EHR system, the CDSS provided real-time recommendations based on patient data. Training sessions for healthcare providers ensured proper usage and interpretation of the system
Outcomes:

- Significant reduction in medication errors
- Increased adherence to guidelines
- Improved patient safety and outcomes
- Streamlined medication management, saving time and reducing cognitive burden

Case Study 2: Telemedicine for Rural Healthcare Access [13]

In a rural healthcare facility, telemedicine was implemented to enhance access to medical services. The technology included video conferencing and digital health platforms, supported by provider training.

Table 2.2 Case Studies of Smart Computing Technologies in Healthcare

Case Study	*Problem Addressed*	*Technology Utilized*	*Outcomes Achieved*
1.	Implementation of clinical decision support system (CDSS)	Clinical decision support, medication errors, adherence to guidelines	Advanced algorithms, evidence-based guidelines
2.	Remote monitoring for chronic disease management	Chronic disease management, hospital readmissions	Wearable devices, remote monitoring technology
3.	AI-assisted diagnosis in radiology	Lung cancer diagnosis, accuracy of radiological interpretation	AI, image recognition system
4.	Telemedicine for rural healthcare access	Healthcare access, geographical barriers to care	Telemedicine platforms, telecommunications technology

Outcomes:

- Improved access to healthcare for remote residents
- Reduced healthcare costs by eliminating travel
- Enhanced patient outcomes and satisfaction

These cases highlight the transformative potential of smart computing in optimizing clinical decision-making and improving healthcare accessibility. Table 2.2 provides a concise summary of the case studies, outlining the specific problems addressed, the technologies utilized to solve them, and the outcomes achieved.

2.4 INTEGRATION OF COGNITIVE AND SMART COMPUTING IN HEALTHCARE

The convergence of cognitive computing and smart technologies is revolutionizing healthcare by enhancing patient care, streamlining operations, and driving innovation. This chapter explores the benefits, challenges, and strategies for successful implementation of this integration. Figure 2.1 demonstrates the integration of cognitive and smart computing in healthcare. It shows data flowing from smart devices to cognitive computing systems, which analyze and provide insights to healthcare providers. These insights inform decision-making, leading to improved patient care.

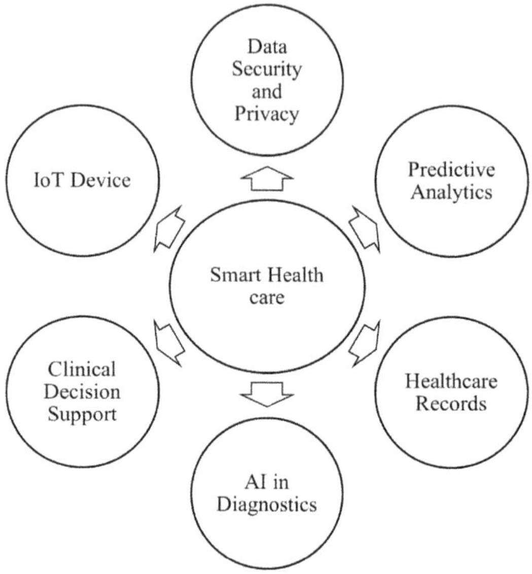

Figure 2.1 Cognitive-Smart Healthcare: Data Flow Dynamics.

2.4.1 Synergistic Benefits of Combining Cognitive and Smart Computing

Integrating cognitive and smart computing in healthcare offers numerous benefits:

1. **Improved diagnostic accuracy:** AI-powered cognitive systems analyze medical data with high accuracy. Smart technologies like imaging devices enhance diagnostic capabilities, leading to earlier disease detection and precise treatment recommendations. AI imaging analysis assists radiologists in identifying abnormalities, improving accuracy and reducing misdiagnosis.
2. **Personalized treatment plans:** Cognitive systems analyze medical history, genetics, lifestyle, and biometric data to create personalized treatments. Data from wearable devices and sensors help deliver targeted interventions, improving treatment efficacy and patient satisfaction.
3. **Efficient healthcare delivery:** AI chatbots and virtual assistants automate administrative tasks, freeing up staff for patient care. Smart algorithms optimize patient flow and resource use, reducing wait times and increasing efficiency.
4. **Continuous monitoring and early intervention:** Combining cognitive systems with wearables and IoT sensors enables real-time health

monitoring. Analyzing biometric data helps detect health deterioration at an earlier stage, allowing for timely interventions and reducing hospital readmissions [14].

5. **Data-driven research and innovation:** Integrating cognitive and smart computing enables large-scale healthcare data analysis, driving medical research and innovation. AI models forecast disease outbreaks, assess treatments, and optimize clinical trials, advancing medical discovery.

2.4.2 Challenges and Limitations

Despite the benefits, integrating cognitive and smart computing into healthcare presents challenges:

1. **Data privacy and security:** Access to vast amounts of sensitive patient data raises concerns about privacy and security. Healthcare organizations must implement robust security measures like encryption, access controls, and regular audits to protect patient information from unauthorized access and cyberattacks.
2. **Interoperability:** Healthcare data is often fragmented across different systems, hindering seamless information exchange. Achieving interoperability requires overcoming technical and organizational barriers. Investment in standards like Fast Healthcare Interoperability Resources (FHIR) and robust data integration frameworks are essential for effective data communication and security.
3. **Ethical and regulatory considerations:** AI in healthcare raises ethical and regulatory issues regarding accountability, transparency, and bias. ML algorithms may perpetuate biases, so organizations must ensure fairness, transparency, and accountability. This includes conducting algorithmic audits and addressing bias concerns. Compliance with laws like the Health Insurance Portability and Accountability Act (HIPAA) and the General Data Protection Regulation (GDPR) are also crucial.

Additionally, the deployment of ethical AI and obtaining patient consent are critical issues. Ensuring ethical AI involves several key components:

1. **Transparency:** Developing transparent AI algorithms where the decision-making process is understandable to both healthcare professionals and patients. This can help in building trust and ensuring that AI decisions can be scrutinized and validated.
2. **Bias mitigation:** Implementing strategies to detect and reduce biases in AI algorithms. This includes using diverse datasets during training and regularly auditing AI systems to ensure they do not perpetuate or amplify existing biases in healthcare.

3. **Accountability:** Establishing clear lines of accountability for AI-driven decisions. This means that healthcare providers and AI developers must take responsibility for the outcomes of AI systems, ensuring that there is a mechanism for addressing any errors or adverse effects.

Patient consent is equally important when deploying AI in healthcare:

1. **Informed consent:** Patients must be fully informed about how their data will be used by AI systems. This includes explaining the benefits, risks, and limitations of AI-driven healthcare solutions in a clear and understandable manner.
2. **Autonomy and control:** Ensuring that patients have control over their data and the ability to opt in or out of AI-based treatments and data sharing. This respects patient autonomy and fosters a sense of trust and collaboration between patients and healthcare providers.
3. **Data privacy:** Protecting patient data through stringent privacy measures. This involves not only securing data but also ensuring that patients' data is used ethically and responsibly, with their consent.

Emphasizing the responsible use of AI is crucial to improve patient outcomes and enhance the healthcare experience. Healthcare organizations must commit to leveraging AI technologies in ways that are ethical, transparent, and beneficial to patients. This includes continuous monitoring and evaluation of AI systems to ensure they are performing as intended and contributing positively to patient care. Addressing these challenges requires coordinated efforts from healthcare organizations, technology vendors, policymakers, and regulatory agencies. By ensuring robust security, promoting interoperability, and adhering to ethical guidelines, healthcare organizations can leverage cognitive and smart computing while protecting patient privacy and data security.

2.4.3 Strategies for Successful Implementation

To overcome challenges in integrating cognitive and smart computing in healthcare, organizations can adopt the following strategies:

1. **Collaboration and partnerships:** Establish partnerships with technology vendors, research institutions, and regulatory agencies to leverage diverse expertise and resources. These collaborations foster knowledge sharing, facilitate technology adoption, and promote the development of interoperable solutions aligned with industry best practices.
2. **Data governance frameworks:** Implement robust data governance frameworks to ensure ethical and responsible data use. Establish

policies for data collection, storage, access, and sharing that comply with regulations like HIPAA and GDPR. Define clear roles, enforce security and privacy controls, and conduct regular audits to protect patient information [15].
3. **Education and training:** Provide specialized training for healthcare professionals to effectively use cognitive computing systems and interpret their insights. Invest in comprehensive training programs covering data literacy, AI fundamentals, clinical decision support, workflow integration, and patient communication. Offer ongoing education to keep professionals updated on emerging technologies and best practices.
4. **Continuous evaluation and improvement:** Establish monitoring and evaluation processes to assess the performance and outcomes of cognitive computing systems. Define key performance indicators (KPIs) to measure accuracy, efficiency, usability, and user satisfaction. Collect and analyze user feedback, stakeholder input, and clinical outcomes data to identify areas for improvement and ensure systems remain aligned with user needs and workflows [16].

Implementing these strategies will help healthcare organizations effectively integrate cognitive and smart computing technologies, improving patient care, enhancing clinical decision-making, and driving innovation in healthcare delivery. Table 2.3 summarizes the synergistic benefits, challenges, and strategies for integrating cognitive and smart computing in healthcare.

2.5 IMPACT ON PATIENT CARE

Technological advancements have significantly transformed patient care, improving outcomes, enhancing engagement, and fostering personalized medicine.

2.5.1 Improved Patient Outcomes

Technological innovations have enhanced diagnosis, treatment, and patient outcomes across medical conditions:

1. **Precision medicine techniques:** Utilize genomics, proteomics, and metabolomics for tailored treatments based on a patient's genetic makeup. This approach improves treatment efficacy and reduces adverse effects, particularly in oncology and pharmacogenomics [17].
2. **Advanced imaging technologies:** Magnetic resonance imaging (MRI) and computed tomography (CT) scans provide detailed anatomical and functional information, enabling early and accurate diagnosis.

Table 2.3 Integration of Cognitive and Smart Computing in Healthcare

Aspect	*Synergistic Benefits*	*Challenges and Limitations*	*Strategies for Successful Implementation*
Improved decision-making	Enhanced diagnostic accuracy through AI-powered analysis of medical data. Personalized treatment plans based on patient-specific information.	Data privacy and security concerns. Ethical considerations related to algorithmic bias and transparency.	Collaboration with technology vendors, research institutions, and regulatory agencies. Investment in robust security measures and ethical guidelines.
Predictive analytics	Early detection and prevention of adverse events such as chronic conditions and hospital-acquired infections.	Fragmented healthcare data silos hindering data integration. Regulatory compliance requirements for predictive models.	Implementation of interoperability standards such as FHIR. Adherence to regulatory guidelines for data privacy and predictive analytics.
Remote patient monitoring	Continuous monitoring of patient health outside traditional healthcare settings. Timely interventions based on real-time biometric data.	Lack of standardized data formats for IoT devices and wearables. Ensuring patient consent and data transparency.	Development of data governance frameworks for secure data sharing. Education and training programs for healthcare professionals.

Imaging-guided interventions target pathological lesions precisely, improving outcomes and reducing recovery times [18].

3. **Electronic health records (EHRs):** EHRs enhance communication, coordination, and care continuity by providing real-time access to comprehensive patient information. They support decision-making and reduce medical errors, leading to better patient outcomes.

2.5.2 Enhanced Patient Engagement

Digital health technologies powered by smart computing empower patients to actively participate in their healthcare journey:

1. **Patient portals and mobile health applications (apps):** These tools go beyond basic functionalities to enhance patient engagement. AI-powered chatbots can answer questions and schedule appointments, while data visualization tools present complex health data in an understandable format, fostering informed decision-making [19].

2. **Telemedicine platforms:** Smart computing takes telemedicine further. ML algorithms analyze patient data remotely and trigger alerts if needed, while real-time data capture allows for comprehensive remote monitoring and personalized intervention [20].
3. **Wearable devices:** Smart algorithms personalize health goals, provide actionable insights based on real-time data, and offer tailored feedback. Integration with other health data sources creates a holistic view of patient health for personalized care plans [21].
4. **Virtual reality (VR) and augmented reality (AR) technologies:** Adaptive VR/AR training programs adjust difficulty and content based on individual needs, while AI-powered feedback systems offer personalized guidance during VR/AR rehabilitation exercises, improving treatment adherence and outcomes [21].

2.5.3 Personalized Medicine

Technological innovations drive personalized medicine by analyzing large-scale data and tailoring treatments to individual patients:

1. **Smart computing in genomics and proteomics:** Advanced algorithms process genomic and molecular data to identify personalized treatment options, optimizing therapeutic outcomes [22].
2. **Cognitive computing for biomarker identification:** AI and ML analyze biomarker data to stratify patients and enhance treatment strategies.
3. **Machine learning (ML) for predictive analytics:** Predictive analytics identify patient responses to treatments, enabling personalized recommendations.
4. **Point-of-care testing devices (POCT):** Smart computing enables rapid diagnostic testing, allowing real-time, personalized interventions at the point of care [23]. Technological innovations have revolutionized patient care, leading to improved outcomes, enhanced engagement, and the realization of personalized medicine, ultimately improving the quality of life for individuals worldwide. Figure 2.2 illustrates how advanced tools like imaging scans, genetic testing, robotic surgery, and telemedicine platforms seamlessly integrate into key stages from diagnosis to follow-up care.

2.6 IMPACT ON CLINICAL DECISION-MAKING

The integration of AI and ML in healthcare has transformed clinical decision-making. These advancements have significantly improved diagnostic accuracy, treatment planning, and reduced medical errors.

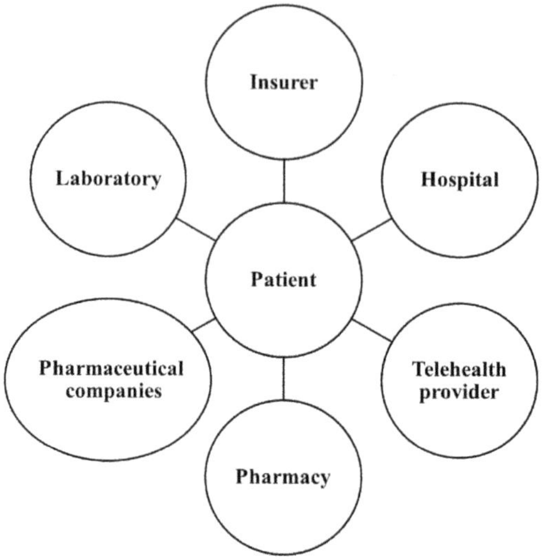

Figure 2.2 Enhancing the Patient Journey: Technology Integration in Healthcare.

AI and ML have revolutionized diagnostic methodologies, enhancing accuracy and efficiency:

1. **Advanced algorithms:** AI leverages sophisticated algorithms to analyze extensive medical datasets, improving precision in diagnosing illnesses. These systems interpret complex data such as symptoms, medical imaging, and lab tests with heightened accuracy.
2. **Deep learning in medical imaging:** AI algorithms, particularly deep learning, excel in analyzing X-rays, MRIs, and CT scans. They identify subtle abnormalities and specific biomarkers, providing clinicians with valuable diagnostic insights and enabling timely, precise decision-making.
3. **Empirical evidence:** Studies show AI algorithms outperform human experts in detecting conditions like malignant tumors from medical images, underscoring their transformative potential in diagnostics [24].
4. **Continuous learning:** AI systems continuously learn and adapt by assimilating new data, enhancing their diagnostic capabilities over time. This dynamic decision support tool provides real-time, evidence-based recommendations, optimizing clinical outcomes. The integration of AI and ML in healthcare marks a paradigm shift, offering unparalleled opportunities to enhance diagnostic accuracy, streamline workflows, and improve patient outcomes, thereby shaping

Table 2.4 Cognitive-Based Smart Computing Diagnostic Accuracy vs. Human Experts

Diagnostic Task	*AI Accuracy (%)*	*Human Expert Accuracy (%)*
Tumor detection	98.5	92
Disease prediction	94.3	86.5
Anomaly identification	99.2	91.8

the future of healthcare delivery. Table 2.4 compares the diagnostic accuracy of cognitive-based smart computing systems with that of human experts across various tasks, including tumor detection, disease prediction, and anomaly identification. The data presented here is hypothetical and for illustrative purposes only. Nonetheless, it showcases the potential superiority of cognitive-based smart computing systems in enhancing diagnostic capabilities and improving patient outcomes.

2.7 ENHANCED TREATMENT PLANNING

Cognitive-based smart computing, powered by AI and ML, has revolutionized treatment planning by offering personalized, evidence-based recommendations. These technologies analyze vast datasets, including patient demographics, medical histories, genetic profiles, and treatment outcomes, to create tailored treatment plans that optimize patient care.

1. **Predictive analytics models:** These models forecast patient responses to various treatments by analyzing clinical parameters, genetic markers, and treatment histories. This predictive capability helps clinicians make informed decisions about interventions, dosage adjustments, or alternative therapies, maximizing treatment efficacy and minimizing adverse effects.
2. **AI-Powered clinical decision support systems (CDSSs):** These systems analyze EHRs in real-time, providing clinicians with clinical guidelines, drug interaction warnings, adverse event alerts, and risk assessments at the point of care. By synthesizing evidence-based knowledge and patient-specific information, these systems enable well-informed, personalized treatment decisions.

Leveraging cognitive-based smart computing insights and predictive analytics, healthcare providers can devise personalized treatment strategies that are more effective, efficient, and aligned with patients' specific needs, improving patient outcomes and optimizing resource utilization.

2.7.1 Reduced Medical Errors

Medical errors significantly threaten patient safety and contribute to adverse outcomes. Cognitive-based smart computing, powered by AI and ML, offers promising solutions to mitigate these risks by identifying potential errors and providing decision support to healthcare professionals. These systems use advanced algorithms and NLP to analyze vast amounts of medical data with precision. NLP algorithms can review clinical notes and documentation to identify inconsistencies, errors, or missed diagnoses, alerting healthcare providers to areas requiring further review or intervention.

By proactively identifying and addressing potential errors, cognitive-based smart computing technologies enhance the overall quality and safety of patient care, contributing to reduced medical errors, better healthcare outcomes, and improved patient safety.

2.8 IMPACT ON RESOURCE ALLOCATION

In modern healthcare systems, the effective allocation of resources is paramount for ensuring quality patient care, maximizing efficiency, and minimizing costs. The integration of technology and data analytics has revolutionized resource allocation, leading to optimized resource utilization, reduced healthcare costs, and improved operational efficiency.

2.8.1 Optimized Resource Utilization

Optimizing resource utilization in healthcare involves efficiently allocating resources such as personnel, equipment, and facilities to meet patient needs while minimizing waste and enhancing overall productivity. Cognition-based smart computing, leveraging AI and ML, plays a pivotal role in achieving this goal by augmenting human decision-making capabilities and enabling proactive resource management based on cognitive insights.

1. **Data-driven decision making (DDDM) enhanced by cognitive computing:** Cognition-based smart computing enhances data-driven decision-making in healthcare by leveraging advanced algorithms to analyze vast datasets encompassing patient volumes, treatment outcomes, and resource utilization patterns. Unlike traditional analytics approaches, cognition-based systems emulate human cognitive processes, enabling deeper understanding and extraction of actionable insights from complex data sources [25].
2. **Dynamic resource allocation empowered by cognitive insights:** Real-time cognitive analytics empower healthcare facilities to dynamically allocate resources based on cognitive insights derived from continuous data streams. By integrating real-time patient data,

environmental factors, and operational metrics, cognition-based systems can anticipate fluctuations in patient demand and proactively adjust staffing levels, bed capacities, and other resources to optimize resource utilization and ensure timely delivery of care. Cognition-based smart computing enables dynamic resource allocation in healthcare by leveraging real-time cognitive insights to anticipate changes in demand and optimize resource utilization without compromising patient care [26].

3. **Enhanced collaboration and resource sharing through cognitive platforms:** Cognition-based platforms facilitate enhanced collaboration and resource sharing among healthcare institutions by providing intelligent interfaces that enable seamless exchange of medical records, diagnostic imaging, and expertise. These platforms leverage cognitive capabilities to interpret and contextualize data, enabling more effective decision-making and resource utilization across diverse healthcare settings. Cognition-based platforms foster collaboration and optimize resource utilization in healthcare by providing intelligent interfaces that facilitate seamless exchange of information and expertise.

Cognition-based smart computing revolutionizes resource utilization in healthcare by augmenting data-driven decision-making, enabling dynamic resource allocation, and fostering enhanced collaboration across healthcare ecosystems. By harnessing the power of cognitive insights, healthcare organizations can optimize resource allocation, improve operational efficiency, and ultimately enhance patient outcomes. Below is a comparison Table 2.5 of resource allocation strategies, contrasting traditional approaches with data-driven approaches.

In staffing, traditional methods rely on historical averages and fixed schedules, often causing inefficiencies during demand fluctuations. In

Table 2.5 Resource Allocation Strategies

Resource Allocation Strategy	*Traditional Approach*	*Data-Driven Approach*
Staffing levels	Fixed schedules based on historical averages	Dynamic scheduling based on real-time demand forecasting.
Equipment utilization	Manual tracking and ad-hoc maintenance	Predictive maintenance schedules based on usage patterns.
Facility management	Static bed assignments and inventory management	Real-time bed management and predictive analytics for capacity planning.

contrast, data-driven approaches use real-time data and predictive analytics to forecast demand, dynamically adjusting staffing to align with current patient needs. This ensures optimal staffing levels, improving patient care and operational efficiency. Regarding equipment, traditional practices involve manual tracking and reactive maintenance, leading to underutilization and downtime. Data-driven approaches employ predictive maintenance based on usage patterns and equipment health data to minimize disruptions, enhance reliability, and reduce maintenance costs in healthcare settings. Facility management transitions from static bed assignments and manual inventory to real-time bed management and predictive analytics optimize resource use and patient flow. Embracing data-driven strategies enhances operational efficiency, improves resource allocation accuracy, and enhances care quality for patients.

2.8.2 Reduced Healthcare Costs

Efficient resource allocation driven by cognition-based smart computing not only enhances patient care but also generates significant cost savings within healthcare systems.

1. **Preventative care and chronic Disease management:** Cognition-based smart computing enables proactive resource allocation towards preventative care and chronic disease management. By leveraging advanced analytics, healthcare providers can identify at-risk populations early and intervene to mitigate chronic conditions. For instance, predictive modelling identifies individuals at high risk of diabetes complications, enabling targeted interventions that reduce costly advanced treatments [27].
2. **Elimination of waste:** Cognitive analysis of large datasets helps healthcare organizations identify inefficiencies and areas of resource waste. By pinpointing underutilized resources and recommending optimizations, cognitive algorithms streamline operations, reduce unnecessary expenditures, and improve resource allocation efficiency [28].
3. **Value-based care models:** Cognition-based smart computing supports value-based care models by optimizing resource allocation strategies. By analyzing patient data and cognitive insights, healthcare providers prioritize interventions that maximize patient outcomes while minimizing costs [29].

2.8.3 Improved Operational Efficiency

Efficient resource allocation enhances operational efficiency across all aspects of healthcare delivery, from patient scheduling to clinical workflows.

1. **Streamlined patient flow:** Cognitive-based smart computing analyzes patient data and operational metrics to optimize patient flow processes. By predicting patient arrivals and allocating resources accordingly, these systems minimize bottlenecks and ensure timely access to care, improving overall patient experience.
2. **Enhanced care coordination:** Integrated platforms enabled by cognitive-based smart computing facilitate seamless communication and care coordination among healthcare providers. Real-time insights ensure timely interventions and reduce treatment delays, enhancing care delivery efficiency and patient outcomes [30].
3. **Staff optimization:** Cognitive-based smart computing empowers healthcare organizations to optimize staffing levels in response to fluctuating patient demand. By dynamically adjusting schedules, these systems enhance workforce satisfaction and operational efficiency, ensuring efficient resource utilization without compromising patient care.

Efficient resource allocation through cognition-based smart computing optimizes operational workflows, reduces costs, and improves patient outcomes, thereby advancing healthcare delivery in a sustainable manner. The core components of the value-based care model, emphasizing its focus on improving healthcare delivery. At the center is Patient-Centered Care, highlighting its central role. Surrounding it are interconnected circles representing the Principles of Value-Based Care: Patient Outcomes, Cost-Effectiveness, and Care Coordination. Each principle is supported by specific strategies and initiatives depicted in the diagram. Additionally, Enablers of Value, Based Care, such as technology and payment models, are positioned around the edges, signifying their role in supporting successful implementation. Overall, Figure 2.3 provides a concise overview of the value-based care model's holistic approach to enhancing patient outcomes and healthcare efficiency.

2.9 IMPACT ON HEALTHCARE WORKFORCE

In healthcare, cognitive-based smart computing technologies have profoundly transformed both patient care and the healthcare workforce. This section explores their tangible impacts on healthcare personnel, emphasizing enhanced productivity, reduced workload, and improved job satisfaction.

1. **Enhanced productivity:** Cognitive-based smart computing systems have revolutionized productivity in healthcare by automating tasks like data entry, documentation, and scheduling. These systems leverage real-time data and predictive analytics to optimize resource

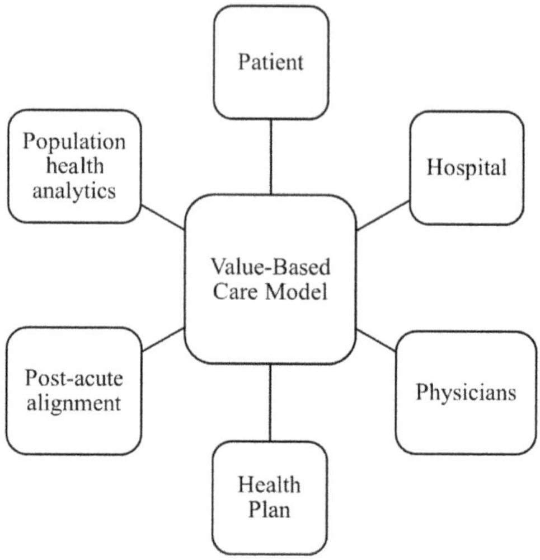

Figure 2.3 Value-Based Care Model: Enhancing Healthcare Delivery.

allocation, freeing up healthcare professionals to focus more on direct patient care. For instance, AI-powered CDSSs aid in diagnosing and treating patients efficiently by synthesizing vast amounts of data and medical knowledge [31].

2. **Reduced workload:** Integration of cognitive computing has significantly lightened the workload of healthcare professionals by automating repetitive tasks and improving resource management. Predictive analytics help hospitals anticipate patient needs and adjust staffing levels, accordingly, ensuring optimal care delivery during peak times. This shift reduces burnout and allows professionals to dedicate more energy to patient interactions and professional growth, enhancing overall efficiency and satisfaction.
3. **Improved job satisfaction:** Cognitive-based smart computing enhances job satisfaction by empowering clinicians with advanced decision support tools and automating administrative tasks. By streamlining workflows and promoting collaboration across specialties, these technologies foster a supportive work environment where professionals can focus on delivering high-quality care and advancing their skills. This holistic approach improves patient outcomes and cultivates a fulfilling work experience [32].

In summary, cognitive-based smart computing is reshaping the healthcare workforce by boosting productivity, reducing workload, and enhancing job satisfaction. By championing efficiency and professional development, these innovations are paving the way for a resilient and fulfilling healthcare environment focused on exceptional patient care.

2.10 CHALLENGES AND LIMITATIONS

In the realm of cognitive-based smart computing, several challenges hinder widespread adoption. This section addresses key concerns: data privacy and security, lack of standardization, and ethical considerations.

1. **Data privacy and security concerns:** Ensuring the confidentiality and integrity of vast data sets is crucial in cognitive computing. Privacy extends beyond personal information to safeguard insights derived from analytics. Encryption and robust authentication are essential to mitigate risks [33].
2. **Lack of standardization and interoperability:** The absence of standardized frameworks complicates integrating diverse hardware and software. This hampers data exchange and scalability, leading to higher costs and compatibility issues [34]. Efforts to establish common standards and APIs aim to improve interoperability, enabling seamless operation across systems. This is a critical challenge in healthcare, as highlighted by studies exploring stakeholder perspectives on interoperability [35].
3. **Ethical considerations:** Ethical dilemmas arise from the autonomous nature of cognitive systems. Algorithmic biases pose risks in decision-making, affecting areas like employment and healthcare. Transparency and fairness in algorithm design are essential to mitigate discriminatory impacts, as highlighted in a recent systematic literature review on cognitive computing trends in healthcare [36,37].

Moreover, ethical data collection and usage, along with transparency in system operations, are critical. Upholding privacy and consent principles ensures responsible deployment and positive societal impact. Addressing these challenges is pivotal for the ethical development and deployment of cognitive-based smart computing, ensuring alignment with societal values and regulatory standards.

2.11 CONCLUSION

Cognitive-based smart computing holds immense potential for revolutionizing healthcare delivery through advancements in artificial intelligence and machine learning. These technologies offer benefits such as enhanced diagnostic accuracy, personalized treatment recommendations, and improved patient outcomes. However, alongside these benefits come significant challenges, including concerns about data privacy, algorithm bias, and the interpretability of machine-generated diagnoses. Looking towards the future, continued innovation in AI algorithms and the integration of cognitive computing with emerging technologies like blockchain and genomics promise further advancements in healthcare. Collaborative efforts among stakeholders are essential to address challenges and establish robust standards for data governance, ethical AI deployment, and patient consent.

In conclusion, cognitive-based smart computing is transforming healthcare delivery, empowering providers to deliver personalized care efficiently. As we navigate this evolving landscape, responsible deployment of these technologies is essential to maximize benefits while ensuring patient privacy and ethical standards are upheld.

REFERENCES

[1] Agency for Healthcare Research and Quality (AHRQ). (n.d.). *National Quality Forum Safe Practices and Standards.* Retrieved from www.ahrq.gov/patient-safety/resources/index.html

[2] Johnson, K. B., Wei, W. Q., Weeraratne, D., Frisse, M. E., Misulis, K., Rhee, K., & Snowdon, J. L. (2021). Precision medicine, AI, and the future of personalized health care. *Clinical and Translational Science, 14*(1), 86–93. https://doi.org/10.1111/cts.12884

[3] Abernethy, A., Adams, L., Barrett, M., Bechtel, C., Brennan, P., Butte, A., & Valdes, K. (2022). The promise of digital health: Then, now, and the future. *NAM Perspectives, 2022*, 10.31478/202206e. https://doi.org/10.31478/202206e

[4] Cowie, M. R., & Lam, C. S. P. (2021). Remote monitoring and digital health tools in CVD management. *Nature Reviews Cardiology, 18*(7), 457–458. https://doi.org/10.1038/s41569-021-00548-x

[5] Hossain, E., Rana, R., Higgins, N., Soar, J., Barua, P. D., Pisani, A. R., & Turner, K. (2023). Natural language processing in electronic health records in relation to healthcare decision-making: A systematic review. *Computers in Biology and Medicine, 155*, 106649. https://doi.org/10.1016/j.compbiomed.2023.106649

[6] Benson, A. B., Venook, A. P., Al-Hawary, M. M., Cederquist, L., Chen, Y. J., Ciombor, K. K., & Robertson, M. (2020). NCCN guidelines insights: Colon cancer, version 2.2020. *Journal of the National Comprehensive Cancer Network, 18*(5), 1–10.

[7] UCSF Center for Digital Health Innovation. (n.d.). *CareWeb.* Retrieved from https://innovation.ucsf.edu/departments/Center%20for%20Digital%20Health%20Innovation

[8] Vijayan, V., Connolly, J. P., Condell, J., McKelvey, N., & Gardiner, P. (2021). Review of wearable devices and data collection considerations for connected health. *Sensors (Basel), 21*(16), 5589. https://doi.org/10.3390/s21165589

[9] Haleem, A., Javaid, M., Singh, R. P., & Suman, R. (2021). Telemedicine for healthcare: Capabilities, features, barriers, and applications. *Sensors International, 2*, 100117. https://doi.org/10.1016/j.sintl.2021.100117

[10] Srivastava, J., Routray, S., Ahmad, S., & Waris, M. M. (2022). Internet of Medical Things (IoMT)-based smart healthcare system: Trends and progress. *Computational Intelligence and Neuroscience, 2022*, 7218113. https://doi.org/10.1155/2022/7218113

[11] Shahmoradi, L., Safdari, R., Ahmadi, H., & Zahmatkeshan, M. (2021). Clinical decision support systems-based interventions to improve medication outcomes: A systematic literature review on features and effects. *Medical Journal of the Islamic Republic of Iran, 35*(27), https://doi.org/10.47176/mjiri.35.27

[12] Leo, D. G., Buckley, B. J. R., Chowdhury, M., Harrison, S. L., Isanejad, M., Lip, G. Y. H., & Lane, D. A. (2022). Interactive remote patient monitoring devices for managing chronic health conditions: Systematic review and meta-analysis. *Journal of Medical Internet Research, 24*(11), e35508. https://doi.org/10.2196/35508

[13] Shanti, H., Raman, R., Chakravartty, S., Belgaumkar, A. P., & Patel, A. G. (2022). Effect of the learning curve on survival after laparoscopic liver resection for colorectal metastases. *BJS Open, 6*(2), zrac020. https://doi.org/10.1093/bjsopen/zrac020

[14] Raskar, D., Iyer, R., & Maralapalle, V. (2025). Quantum data processing and analysis: Revolutionizing data science. In *Advances in biomedical informatics, quantum computing for healthcare data* (pp. 189–214). Academic Press. https://doi.org/10.1016/B978-0-443-29297-2.00008-3

[15] Morley, J., Murphy, L., Mishra, A., Joshi, I., & Karpathakis, K. (2022). Governing data and artificial intelligence for healthcare: Developing an international understanding. *JMIR Formative Research, 6*(1), e31623. https://doi.org/10.2196/31623

[16] An, Q., Rahman, S., Zhou, J., & Kang, J. J. (2023). A comprehensive review on machine learning in healthcare industry: Classification, restrictions, opportunities, and challenges. *Sensors, 23*(4178), 1–20. https://doi.org/10.3390/s23094178

[17] Marques, L., Costa, B., Pereira, M., Silva, A., Santos, J., Saldanha, L., Silva, I., Magalhães, P., Schmidt, S., & Vale, N. (2024). Advancing precision medicine: A review of innovative in silico approaches for drug development, clinical pharmacology, and personalized healthcare. *Pharmaceutics, 16*(332). https://doi.org/10.3390/pharmaceutics16030332

[18] Jiang, X., Hu, Z., Wang, S., & Zhang, Y. (2023). Deep learning for medical image-based cancer diagnosis. *Cancers, 15*(14), 3608. https://doi.org/10.3390/cancers15143608

[19] Iyer, R., Maralapalle, V., Mahesh, P., & Patil, D. (2024). Generative AI and LLM: Case study in e-commerce. In *Generative AI and LLMs: Natural language processing and generative adversarial networks* (pp. 253–272). De Gruyter. https://doi.org/10.1515/9783111425078-013

[20] Burge, K. Y., Gunasekaran, A., Makoni, M. M., Mir, A. M., & Burkhart, H. M. (2022). Clinical characteristics and potential pathogenesis of cardiac necrotizing enterocolitis in neonates with congenital heart disease: A narrative review. *Journal of Clinical Medicine, 11*(14), 3987. https://doi.org/10.3390/jcm11143987

[21] Plaskon, D., Evensen, C., Henderson, K., Palatnik, B., Ishikuri, T., Wang, H. C., ... & Thomas Record, M. Jr. (2022). Step-by-step regulation of productive and abortive transcription initiation by pyrophosphorolysis. *Journal of Molecular Biology, 434*(13), 167621. https://doi.org/10.1016/j.jmb.2022.167621

[22] Fonseca, M. H., Kovaleski, F., Picinin, C. T., Pedroso, B., & Rubbo, P. (2021). E-health practices and technologies: A systematic review from 2014 to 2019. *Healthcare (Basel), 9*(9), 1192. https://doi.org/10.3390/healthcare9091192

[23] Price, C. P. (2001). Point of care testing. *BMJ, 322*(7297), 1285–1288. https://doi.org/10.1136/bmj.322.7297.1285

[24] Harmon, S. A., Sanford, T. H., Xu, S., Turkbey, E. B., Roth, H., Xu, Z., ... & Turkbey, B. (2020). Artificial intelligence for the detection of COVID-19 pneumonia on chest CT using multinational datasets. *Nature Communications, 11*(1), 4080. https://doi.org/10.1038/s41467-020-17971-2

[25] Alowais, S. A., Alghamdi, S. S., Alsuhebany, N., Alqahtani, T., Alshaya, A. I., Almohareb, S. N., ... & Al Harbi, S. (2023). Revolutionizing healthcare: The role of artificial intelligence in clinical practice. *BMC Medical Education, 23*(1), 689. https://doi.org/10.1186/s12909-023-04698-z

[26] Iyer, R., Maralapalle, V. & Mahesh, P. (2025). Virtual reality (VR) and augmented reality (AR) in vocational training. In E. Çela et al. (Eds.), *Transforming vocational education and training using AI* (pp. 73–92). IGI Global. https://doi.org/10.4018/979-8-3693-8252-3.ch004.

[27] Stoumpos, A. I., Kitsios, F., & Talias, M. A. (2023). Digital transformation in healthcare: Technology acceptance and its applications. *International Journal of Environmental Research and Public Health, 20*(4), 3407. https://doi.org/10.3390/ijerph20043407

[28] Dhamanti, I., Nia, I. M., Nagappan, K., & Srikaanth, B. P. (2023). Smart home healthcare for chronic disease management: A scoping review. *Digital Health, 9*, 20552076231218144. https://doi.org/10.1177/20552076231218144

[29] Motlatla, M., & Maluleke, T. X. (2021). Assessment of knowledge about healthcare risk waste management at a tertiary hospital in the northern cape province, south Africa. *International Journal of Environmental Research and Public Health, 18*(2), 449. https://doi.org/10.3390/ijerph18020449

[30] van Engen, V., Bonfrer, I., Ahaus, K. & Buljac-Samardzic, M. (2022). Value-based healthcare from the perspective of the healthcare professional: A systematic literature review. *Frontiers in Public Health, 9*, 800702. https://doi.org/10.3389/fpubh.2021.800702

[31] Vargas, I., Eguiguren, P., Mogollón-Pérez, A. S., Samico, I., Bertolotto, F., López-Vázquez, J., & Vázquez, M. L. (2020). Can care coordination across levels be improved through the implementation of participatory action research interventions? Outcomes and conditions for sustaining changes in five Latin American countries. *BMC Health Services Research, 20*(1), 941. https://doi.org/10.1186/s12913-020-05781-7

[32] Secinaro, S., Calandra, D., Secinaro, A., Muthurangu, V., & Biancone, P. (2021). The role of artificial intelligence in healthcare: A structured literature review. *BMC Medical Informatics and Decision Making*, *21*(1), 125. https://doi.org/10.1186/s12911-021-01488-9

[33] Tursunbayeva, A., & Renkema, M. (2023). "Artificial intelligence in healthcare: Implications for the job design of healthcare professionals. *Asia Pacific Journal of Human Resources*, *61*(4), 845–887. https://doi.org/10.1111/1744-7941.12325

[34] Information Resources Management Association. (2020). *Cognitive Analytics: Concepts, Methodologies, Tools, and Applications*. IGI Global. https://www.igi-global.com/book/cognitive-analytics-concepts-methodologies-tools/23659

[35] Junaid, S. B., Imam, A. A., Balogun, A. O., De Silva, L. C., Surakat, Y. A., Kumar, G., Abdulkarim, M., et al. (2022). Recent advancements in emerging technologies for healthcare management systems: A survey. *Healthcare*, *10*(10), 1940. https://doi.org/10.3390/healthcare10101940

[36] Walker, D. M., Tarver, W. L., Jonnalagadda, P., Ranbom, L., Ford, E. W., & Rahurkar, S. (2023). Perspectives on challenges and opportunities for interoperability: Findings from key informant interviews with stakeholders in Ohio. *JMIR Medical Informatics*, *11*, e43848. https://doi.org/10.2196/43848

[37] Srivani, M., Murugappan, A., & Mala T. Cognitive computing technological trends and future research directions in healthcare—A systematic literature review. *Artificial Intelligence in Medicine*, *138*, 102513. https://doi.org/10.1016/j.artmed.2023.102513

Chapter 3

Cognitive Cybersecurity

Symbolic Deep Learning for Enhanced Decision Support

Anirudh D. Maurya, Ankan Mondal, and Kahkashan Siddavatam

3.1 INTRODUCTION

In today's fast-paced digital world, cybersecurity is a major concern for everyone, from individuals to large organizations and governments. With cyber threats becoming more sophisticated, our traditional ways of staying safe online are no longer enough. That is why we need smarter approaches to help cybersecurity professionals make better decisions.

Modern cybersecurity tools resemble old-school detectives trying to catch high-tech thieves, and as a result, they're struggling to keep up. They fight cyber threats with outdated rules or tricks in number-crunching during their discovery and stopping, while the criminals keep finding smarter ways to strike. It is like catching a slippery fish with a torn net. The result is that sometimes we catch too many false alarms, other times the real dangers get through. Security used to be simpler; now new twists and turns keep emerging. With technology spreading like wildfire, cyber threats are becoming sneakier too. The malicious players always come up with different techniques to pose new cyber threats to systems across the world.

A cyber threat can be defined as an attempt by cybercriminals to obtain any information, violate data integrity constraints, or cause damage to computing devices or networks. The forms of threat include; phishing attempts, malware infections, attacks on Internet of Things (IoT) devices, distributed denial of service attacks, denial of service attacks, spam emails, unauthorized intrusion into networks or mobile devices, fraudulent schemes, and ransomware attacks.

These challenges have increased interest in the integration of Symbolic Deep Learning (SDL) techniques into cybersecurity frameworks. SDL is an amalgamation of symbolic reasoning with deep learning (or DL, which, for the remainder of this chapter, will be called deep learning) that presents a sustainable solution towards the complexities of cybersecurity. By combining the logical inference and abstraction characteristics from symbolic reasoning with the pattern recognition power of deep learning, this system

DOI: 10.1201/9781003501220-3

will be very adept at decision support in cybersecurity [1]. Probably the most important advantages of incorporating symbolic reasoning in deep learning models lie in their increased explainability. Traditionally, deep learning models have acted like black boxes wherein it was tough to understand the reason for a decision. Inclusion of symbolic reasoning will hence allow a cyber security model to provide insights into decision-making processes and increase transparency, hence nurturing trust, in automated systems [2].

Moreover, SDL can make use of domain-specific knowledge in cybersecurity models. This will make the models more adaptive to the fine lines of different threat landscapes through the capturing of expert knowledge in symbolic form. Therefore, one gets more contextually relevant threat detection and response strategies. For this reason, domain-specific knowledge leveraging is important for fighting emerging threats [3].

Another important benefit of SDL includes enhanced cybersecurity defenses that enhance resilience to malicious attacks. Models can use the symbolic reasoning capacity to identify inconsistencies and anomalies in data patterns, hence better detecting and mitigation of highly sophisticated evasion techniques used by malicious actors. This added robustness is integral to keeping cybersecurity systems intact against a backdrop of fast-evolving threats.

In essence, the potential for SDL is to further increase cybersecurity capabilities in the near future. This can be realized when symbolic reasoning and deep learning are leveraged synergistically to build more resilient and robust decision support systems against rapidly evolving threats in today's fastmoving world. Embracing these approaches may thus turn out to be very critical in trying to get ahead of these enemies and protect our digital assets in this radically changing, hyper-connected world.

3.2 PHASES OF CYBERSECURITY ARCHITECTURE

In the field of cyber SDL, the three phases of cybersecurity architecture are crucial for its success.

3.2.1 Phase 1: Development

The initial phase of cyber SDL involves laying the groundwork for the system. This includes defining how we represent knowledge symbolically, such as through rules, logic, and ontologies. Emphasis is placed on creating policies, standards, and best practices not only for traditional security measures but also for integrating symbolic reasoning into deep learning models.

3.2.2 Phase 2: Implementation

The second phase is the time to concretize ideas to practical solutions. Integration in cyber SDL will mean the fusion of deep learning techniques from symbolic artificial intelligence (AI), which will result in hybrid models scrutinizing complex cybersecurity tasks. In this phase, the implementation of deep learning architectures for different tasks such as anomaly detection, malware analysis, and intrusion detection will be combined with symbolic reasoning engines to interpret and reason over the symbolically represented knowledge.

3.2.3 Phase 3: Continuous Monitoring

This phase is important to learn how well the recreated models work in real-world cybersecurity environments. It includes the following; (1) after integration, the behavior of the system will be continuously monitored; (2) the alerts and anomalies detected by the deep learning components of the system will be analyzed. Finally, (3) interpretation of results through symbolic reasoning will be done. In addition to feedback loops, these models will be further enhanced over time based on observed performance and the evolving threat landscape.

SDL models could change everything when it comes to the identification and classification of cyber threats. One could envision this as supercharged detectives blending logical thinking with deep learning smarts. They have developed strong capabilities in detecting indicators of cyberattacks, whether, sneaky network break-ins, malware sneak-ins, or phishing schemes. Not only that, but even better is the knowledge and reassurance that they would not raise any false alarms while performing their role or executing their duties.

That is not all. These models are used for intrusion prevention, too, in case a hacker wants to get past them. By identifying unusual patterns in the data and detecting when something just does not quite add up, bots can sniff out malicious activity. Even when hackers try dodging by using distractions

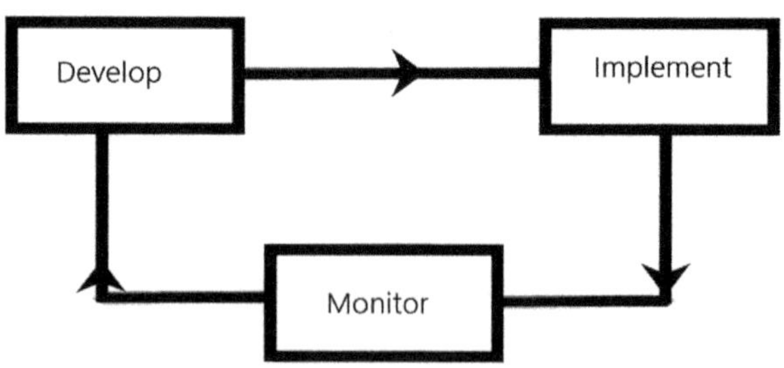

Figure 3.1 Cybersecurity Architecture.

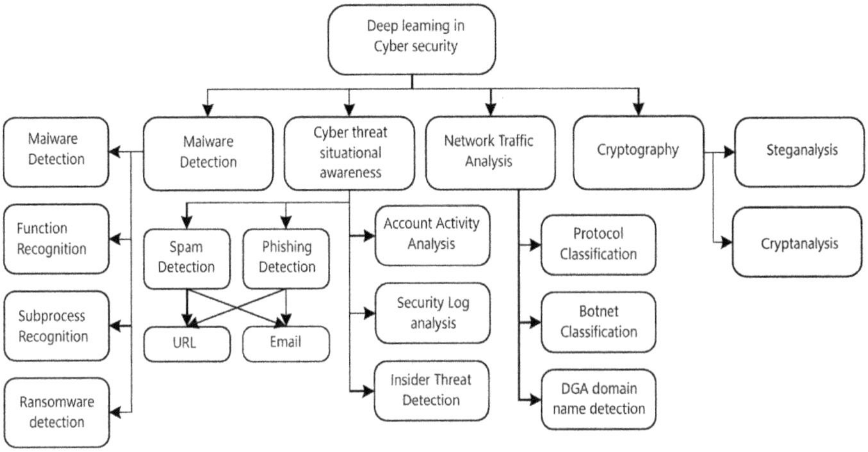

Figure 3.2 Symbolic Deep Learning Model for Cyber Security Applications.

and misdirection, these models can potentially detect attacks as they happen while also reducing false alarms.

3.3 PROBLEM STATEMENT

Cybersecurity becomes increasingly challenging because of the ever-evolving cyber threat landscape. Traditional defense techniques most often fail to detect such threats, leaving gaps in defending against critical infrastructures, sensitive data, and digital assets. Another challenge posed to the problem of defense is the rising complexity and sophistication of cyberattacks. The major problems in the cybersecurity domain include the low effectiveness of existing decision support systems. These systems are typically rule-based or based on statistical techniques, which can no longer keep up with modern hazards. They may also generate very high levels of false positives or miss completely new threats, thus creating inefficiency related to resource allocation and response coordination.

In fact, the existing conventional mechanisms of deep learning models are highly imprecise and ambiguous, acting as the principal line of hindrance in cybersecurity. While deep learning is very good at identifying complex patterns and making accurate predictions, it is still obscure to a great extent; due to that, its inner decision-making processes are hard to understand and, therefore, trust. To that end, the opacity of such models really does dim their practical utility in cybersecurity, in which explanation of decisions is primary for the performance of effective threat analysis and response. On that note, there is a critical need for new approaches that can offer improved decision support for cyber security.

These approaches should address the following key challenges:

1. **Accuracy:** Higher precision and reliability of threat detection to bring down false positives and false negatives.
2. **Explainability:** Decision support systems are expected to become transparent and understandable for cybersecurity professionals to instill trust.
3. **Adaptability:** Develop systems that are adaptable to the dynamic nature of cyber threats; able to predict new attack vectors by proactive defense methods.

In response to these challenges, new approaches should integrate the power of deep learning with symbolic reasoning techniques to improve both interpretability and decision-making. Such approaches will further enhance the ability of cybersecurity professionals to counter advanced cyber threats and defend many other critical digital resources and infrastructures.

Before SDL became the prevalent approach, various methods were in use and applied in cybersecurity decision-making with different strengths and weaknesses, as previously discussed.

One of the earlier methods in use was that of rule-based systems. The systems have pre-defined rules or signatures that identify cyber threats and stipulate how to respond to them. However, usually they are inflexible, not able to respond to new or evolving threats without human intervention. Moreover, they mostly raise a lot of false alarms and have been observed to struggle with efficiently processing large data volumes [4].

Another very pervasive approach in cybersecurity decision-making is the use of statistical methods. It works by looking at historical data for trends or outliers, which otherwise might indicate a possible cyber threat. While statistical methods are useful in a number of domains, they may end up classifying benign activities as threats, therefore leading to high false positive rates. In addition, due to the blindness to deeper relationships between events, such methods have limited insightfulness [5].

Machine learning (ML) techniques have been widely used as well. These methods utilize algorithms that allow them to learn from data on their own to identify potential dangers. However, traditional ML models can sometimes be like black boxes, making it difficult for cybersecurity professionals to understand how they make decisions. This opacity may further make it quite challenging to build trust in this kind of model and hold it responsible for the kind of actions it undertakes [6].

Deep learning as a subdomain of ML has huge potential in cybersecurity because of its ability to learn complex patterns in data. Deep learning models often become black boxes due to the complexity of their architecture whose decisions are not easy to explain or interpret. This lack of transparency thus brings challenges to cybersecurity experts who need explanations for a model's output to build trust in those outputs [7].

These challenges have resulted in an exciting approach to cybersecurity decision-making which has now been introduced in this article as symbolic deep learning, or SDL in short.

3.4 RESEARCH OBJECTIVE

Guided by the overall goal of developing more resilient cybersecurity frameworks, this publication aims to determine how SDL can enhance Gatton cybersecurity. Gatton Cybersecurity refers to the cybersecurity approach and expertise developed within the Gatton College of Business & Economics at the University of Kentucky. This perspective uniquely combines business acumen with cybersecurity principles, emphasizing how organizations can effectively manage cyber risks while maintaining operational excellence and strategic business objectives. We target achieving these objectives, finding new ways to deploy SDL for wiser, more explainable cybersecurity tools; trying new models to improve threat detection by filtering out noise and false alerts; developing techniques to explain decisions made by our models for better trust and understanding by cybersecurity experts; predicting and preventing cyber threats before they happen; and finally, testing this new approach in the real world to show it can make a real difference in keeping us safe online.

3.5 LITERATURE REVIEW

3.5.1 Challenges in Cybersecurity Decision-Making

Facing the challenges of cybersecurity decision-making is like navigating a constantly shifting maze, with cunning adversaries lurking around every corner. Traditional methods often struggle to keep up with these dynamic challenges, leaving vulnerabilities open to cyberattacks.

With the growth of cybercrimes, researchers have focused on detection of invasion in the network [8]. In the past, cyberattacks were mainly for conventional computer networks and personal computers. But now, the scope has widened to include communication systems like IoT [9], network connecting vehicles [10], medical IoT setups [11], and advanced 5G networks [12]. These new targets are increasingly vulnerable to cyberattacks. Old defense methods such as antivirus and firewalls software have difficulty dealing with new threats [13]. For this reason, there is a growing trend to use deep learning (which, as a reminder to the read, is the full form of DL) and ML algorithms for network intrusion detection [14–16]. Additionally, researchers are studying algorithms that can work separately or together. Group strategies, a type of ML approach, is gaining popularity for its intrusion detection capabilities and problem prediction ability.

Then there is the challenge of balancing security with usability and privacy. It has proven to be a delicate dance, if we focus too much on security, we might sacrifice usability or privacy, and vice versa. Finding the right balance is tough but essential. In the past, we have tried different approaches to cybersecurity decision-making, but each has its drawbacks. Rule-based systems are rigid and struggle to adapt, statistical methods generate too many false alarms, and deep learning can feel like a mystery box.

That is where explainable decision support systems come in. They are like friendly guides that show us how they reached each decision, building trust and understanding. This transparency is crucial for meeting regulations and improving our cybersecurity expertise.

Looking forward, we need to focus on refining techniques that make our decision support systems more transparent and understandable. By doing so, we can strengthen our defenses against cyber threats, ensuring a safer digital landscape for everyone involved.

3.5.2 Prior Approaches, Limitations, and the Need for Explainable Decision Support

In the world of cybersecurity, decision-making has used various methods over time, each with its own strengths and weaknesses. Imagine it like having different teachers in school. Some are strict and have stick rules, while others are good at spotting patterns, and then there are those who make teaching look effortless, they seem almost magical, yet never explain how they do it.

The authors in [17] explored how commonly used ML methods can be applied to safeguard cyberspace from cybercriminals. They also discussed the challenges encountered when implementing these techniques. Their findings highlighted the ongoing need for significant advancements to secure classifiers from adversarial attacks. It was concluded that, while ML methods offer promising avenues for cyber defense, there remains a critical need for further development to shield classifiers from cyber threats and attacks.

The authors in [18] provided a concise overview of various studies focusing on using ML models to improve cybersecurity. They discussed the common challenges encountered when applying ML techniques, particularly in identifying suitable datasets with optimal relevance for specific security issues.

The authors in [19] introduced a statistical semi-supervised ML method designed for detecting intrusions on Android mobile devices. As data traffic continues to grow, the risk of cybercrimes also escalates. Therefore, there is a demand for more sophisticated ML techniques to safeguard Android mobile devices from advanced cyber threats by identifying malicious activities effectively [20].

This publication is here to break down the complexities of deep learning and its application in cybersecurity, especially for those eager to understand this field. It serves as a guidebook, tailored for aspiring students who want to use deep learning methods in securing cyber systems.

But now, we have something new on the horizon; explainable decision support systems. These are like friendly guides who walk users through their thought processes, explaining why they make each decision. This transparency not only helps us trust the system more but also makes it easier for us to understand and learn from it. This is crucial for meeting regulations and improving our skills in cybersecurity.

Moving forward, it is essential to focus on improving these systems to make them even more transparent and understandable. By doing so, we can strengthen our defenses against cyber threats and create a safer online world for everyone.

3.6 METHODOLOGY: SYMBOLIC DEEP LEARNING

3.6.1 Applications to Defense Modeling

SDL, which, to remind the reader, is short for symbolic deep learning. joins deep learning with symbolic reasoning, representing a tremendous advance in cybersecurity defense modeling. Think of SDL as your genius colleague; it can detect complicated patterns and explain how it found them.

SDL functions as a sophisticated recipe. It takes the pattern quizzing of deep learning and clarity of symbolic reasoning, quickly improving cybersecurity practices by fully understanding data and making interventions.

In cybersecurity, it is a game changer due to the fact that SDL is able to explain its decisions. It detects the patterns, understands anomalies. This makes it easier for security analysts to understand and reduce risks. SDL stands out as a robust approach to reinforcing cybersecurity defenses.

3.6.1.1 Contributions of SDL in Defense Modeling

It offers a number of benefits in the domain of cyber security. One of SDL's main applications is threat identification and categorization; it scans network logs for suspicious activity. Unlike traditional methods, SDL does more than raising a flag; it explains to the administrators the reason an activity is suspicious. This improves understanding and helps in decision-making.

It is outstanding in anomaly detection, whereby it detects abnormal patterns or behaviors within the cybersecurity data. Not only that, but it also explains the reason for these deviations. This shows its importance in detecting unauthorized access or even anomalous network activities.

Another way SDL contributes to cybersecurity is proactive defense. SDL analyzes historical attack data and helps to anticipate where the threat landscape will go next. The technology analyzes past data or incidents to learn how to respond to potential future vulnerabilities. This standby allows organizations to be well-equipped with the necessary means to fight emerging threats and stay one step ahead.

3.7 MODEL TRAINING

Model training is the critical phase where cybersecurity systems learn from real security data, including system logs, historical attack patterns, and threat intelligence reports. Through analyzing these patterns, the system becomes more adept at distinguishing between legitimate activities and malicious behavior, minimizing false alerts while improving its ability to predict emerging threats. Once training is complete, the system operates with greater reliability and enables organizations to implement more effective proactive defenses.

3.7.1 Step 1: Collection of Data and its Preparation

The model training process begins by gathering huge data from multiple sources to build a large dataset. This data could be taken from system events, network logs, threat reports, historical records of cyberattacks, internal networks of a company, tools of cybersecurity, public databases, simulated scenarios of cyberattacks, user behavior analytics, threat feeds coming in from an external source, dark web monitoring, cloud service logs, IoT device logs, and application logs. The cleaning process sanitizes the data from any irrelevant or duplicate data, normalizes formats, or ensures consistency across different sources. This preprocessing is very important as it minimizes noise and remains with a relevant set of information for training the SDL model.

3.7.2 Step 2: Training the Model with GNN

Once the data is prepared, training of the SDL model begins. It is deep learning that integrates symbolic reasoning to interpret and analyze the data efficiently. Precisely, it boils down to incorporating graph neural networks (GNN) into this very SDL framework. Raw cybersecurity data becomes graph-structured in which the nodes represent entities and edges stand for relationships or interactions, thus capturing most of the complex dependencies and intricate interactions within data. Subsequently, GNN models learn patterns of graph structures to recognize more complex dependencies that are not a priority evident in the classical representations. These kinds of insights learned during GNN training shall be translated into symbolic rules and patterns, which enables translation of learning based on graph-based insights into a format capable of explaining and interpreting the relationships and patterns found by the GNN. These symbolic patterns are leveraged to power and advance the decision-making process by which an organization has access into its' complete cybersecurity data, enabling better threat detection and response strategies.

3.7.3 Step 3: Evaluating SDL Performance

In our experimentation with SDL, utilizing GNN for cybersecurity, the returned model accuracy was 93.41%. The estimation was done using k-fold cross-validation to ensure its robustness. K-fold cross-validation divides the dataset into k equal-sized segments, then trains the model on k-1 segments while testing on the remaining segment. This process repeats k times, ensuring each segment serves as the test set exactly once. This approach provides a robust estimate of how the model will perform on new data and helps prevent overfitting. The average accuracy from the multiple folds is taken.

Testing on a graph constructed from network logs has shown the effectiveness of GNN in handling complex relationships within cybersecurity data. Notably, test results may vary based on the number of features chosen. That is what makes SDL with GNN very promising in providing huge improvement in cybersecurity defenses by placing deep learning pattern recognition with interpretability from symbolic reasoning.

3.7.4 Step 4: Operationalization and Ethical Considerations

After training and testing, it will be integrated into cybersecurity systems to analyze data in real time. Of course, it is monitored and updated, at all times, in the quest to have a model that answers relevant and emerging threats. Extensive testing using new data is done to test the accuracy and robustness of the model toward detecting threats. The model will also be fine-tuned based on feedback and performance metrics for increased accuracy and reliability. In making and using this model, a lot of ethical values have to be observed, including obedience to the law, protection of privacy, and fairness and no discrimination. Ethical considerations should be factored as a whole along the development and implementation of the SDL models in ensuring that tools developed promote a secure and fair digital environment.

Major tasks in the construction of better defenses should then include detailed data collection, serious training processes, and integration of GNNs that provide human-readable inferences. Equipped with GNNs, the SDL framework is then a strong and interpretable strategy for developing effective cybersecurity solutions that will greatly improve the safety of the digital world.

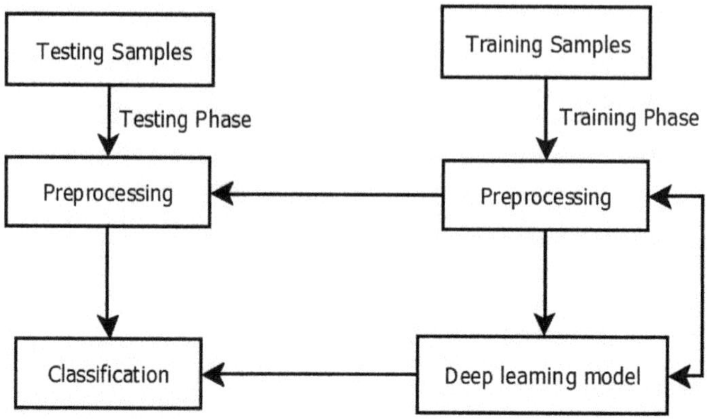

Figure 3.3 Model Training and Testing.

3.8 ASPECTS OF CYBERSECURITY

Different areas of SDL expand across multiple domains in cybersecurity, providing novel and enhanced capabilities. SDL offers interpretable models that find many uses within the domains of cybersecurity. The following sections examine in detail how SDL would change these critical areas.

1. **Endpoint security:** Protecting devices connected to the network is important and should not be overlooked. SDL enables endpoint security to provide transparent and interpretable models so that threats can be identified and mitigated down to the device level, thus refusing entry points from being exploited.
2. **Data security:** Since data is a critical asset, its confidentiality, integrity, and availability/accessibility must be maintained. This may change the techniques used in organizational data protection mechanisms like encryption and masking when SDL comes onboard. On other fronts, SDL will offer models interpretable for the detection of vulnerabilities, among others, and for providing evidence regarding the right application of encryption protocols and their proper maintenance.
3. **Threat intelligence:** SDL makes it easier to collect, analyze, and act on information that describes or indicates a potential or actual threat. By adopting SDL, cybersecurity professionals would get a much clearer view of threat patterns and behaviors, which in most cases helps one to independently protect against cyberattacks by giving more accurate and transparent background/information on the decision-making process.

4. **Incident response:** This can basically improve the development and implementation of procedures for detecting, responding, and recovering from cybersecurity incidents in the SDL. Interpretable SDLs will let one identify the incidents faster; with clear steps on how to act in response and ensure maximum limitation of damage by efficiently restoring normal operations.
5. **Cloud security:** It will help in data breaches, loss of data, and hijacking of accounts. SDL models provide transparency on security measures and would hence allow better monitoring and response in cloud environments against such threats.
6. **Identity and access management (IAM):** Secure access management and access to all organizational resources by user identities are thus enhanced with SDL. SDL can provide transparent, interpretable models for access control dealing explicitly with the principles of least privilege and separation of duties, giving access to critical systems and data only to those authorized, therefore better detecting and preventing unauthorized access.
7. **Network security:** The confidentiality, integrity, and availability/accessibility of data are crucial while it is in transmission across or received by networks. SDL is helpful in modeling and analyzing network traffic to provide interpretable insights that aid in the identification of threats and mitigating them to ensure secure data transmission. It can search for patterns in data to find if any kind of tampering is done to the data.
8. **Endpoint security:** Secure single endpoint devices that connect to a network. SDL can enhance endpoint security through transparent and interpretable models for the detection and mitigation of threats at the device level, thus preventing the exploitation of entry points.

3.9 SYSTEM ARCHITECTURE: SDL FOR MODELING EXPERT BEHAVIOR

3.9.1 Cognitive Models and Transparency

Integrating SDL intelligently enhances befitting digital defenses in cybersecurity systems with transparency. The system learns from top professionals at cybersecurity firms by studying how they handle threats and incidents, then adapting continuously to new challenges and feedback. The system gives clear and understandable explanations of its decisions in the same way that a teacher may break down a complex topic into incredibly simple lessons catering to a student's learning style. It ingests (i.e., receives or collects) data from various sources, like network logs and threat reports, making sure that the data is cleaned/neutralized (i.e., metaphorically, disinfected or decontaminated) and relevant to the task it has been obtained to perform before analysis. Coupling advanced models with this data allows for smarter, more transparent, real-time insights and support. Acting as a trusted mentor, it guides professionals to make informed decisions and thus improves their defense strategies. Human–AI collaboration enables digital defenses to remain adaptive and robust, ensuring they stay ahead of emerging threats.

3.10 EXTRACTING SYMBOLIC KNOWLEDGE

Graph construction is the process of representing raw cyber security data in graphs. These complex relationships and interactions are represented in network security. Sources of such raw data can be network logs, user access records, threat reports, and so on. Nodes in such graphs represent users, devices, Internet Protocol (IP) addresses, and other entities. While an IP address is the unique identifier for a device on some network, nodes represent these different entities. The edges in the graph denote interactions; this could be due to login attempts or data transfers. For example, data from a corporate network may form nodes for IP addresses and devices, while edges will denote access attempts.

1. **Graph neural networks (GNN) training:** Training GNNs involves the use of GNNs as a model, which learns patterns and relationships within this graph data by employing Graph Convolutional Networks (GCNs) or Graph Attention Networks (GATs). In other words, a GNNs involves using architectures such as GCNs or GATs to learn patterns and relationships within graph-structured data. During training, graphs are constructed from log data, and the training dataset includes labeled examples that represent both normal behavior and different types of anomalies (i.e., the training process uses graphs built from log data, where labeled examples include instances of normal activity and various classes of anomalous behavior). The GNN models learn the regularity of, say, normal login attempts versus the unusual activity of potentially threatening behavior. For instance, it could be an anomalous logging-in activity wherein a GNN identifies late-night access to back-end systems emanating from a newly acquired IP address.
2. **Symbolic extraction:** Symbolic extraction is a process in which insights captured by GNNs are interpreted in symbolic rules and patterns. For example, if GNN insights uncover that a user logging in from multiple geographically faraway locations within a very short period of time, is involved in a great number of credential stuffing attacks, that pattern gets translated to, "If a user logs in from locations more

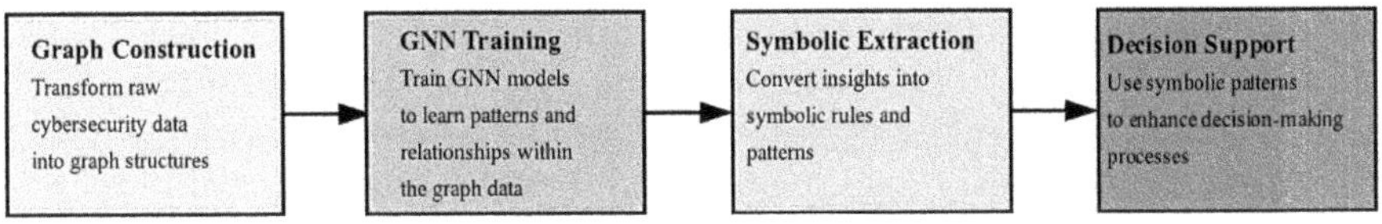

Figure 3.4 Classification Process Model Training.

than 500 miles apart within 24 hours, flag the activity as suspicious." These rules provide clear explanations for GNN findings, and thus complex relationships within a graph are interpretable.

3. **Decision support:** The final stage makes use of symbolic patterns to enhance decision-making processes. The symbolic rules, when applied, generate alerts in real-time for assessing the risks. For instance, one of the symbolic rules might read, "In case of a user account having more than five failed login attempts within an hour and then accessing a file tagged as 'confidential,' flag as 'Potential Brute Force Attack'." An alert stemming from this rule would then say, "Warning: A lot of failed login attempts from the IP address X and then a successful log-in. It may be a brute force attack. Verify immediately." The feedback from these real-world scenarios refines GNN models and symbolic extraction rules for betterment in the accuracy and reliability of the overall system.

Our SDL-based system integrates processes for robust, adaptive, and transparent cybersecurity that is learned and evolved to protect digital environments.

3.11 IMPLICATIONS AND APPLICATIONS OF SYMBOLIC DEEP LEARNING

3.11.1 Bridging Expertise Gaps in Cybersecurity

SDL holds great potential for bridging the gap in expertise towards increasing cyber-resilience. It tends to democratize access to sophisticated decision support systems and empowers a broader range of stakeholders to make informed decisions. Traditional cybersecurity decision support systems require specialist expertise in data analysis, ML, and domains of cybersecurity. This enhanced capability, emphasizing interpretability and transparency; SDL opens up to a larger section of expertise levels among cyber security professionals. It can provide interpretable insight and explanations with which even non-expert users can understand and act on the recommendations.

It is in this light that SDL can also fit very well into cybersecurity education and training, where students can be trained using state-of-the-art hands-on decision support technologies. By including SDL techniques in courses, schools enable the transfer of skills, thus equipping students with the ability to analyze cybersecurity data, understand model predictions, and hence to make informed decisions, arming the next generation of professionals in the field of cybersecurity with proficient ways to ably fight back against fast-evolving cyber threats.

SDL makes collaborative decision-making possible by rendering cybersecurity data and model predictions comprehensible and in an easy-to-understand and insightful format. Cybersecurity professionals from diverse backgrounds will be able to collaborate on SDL-based decision support systems more effectively in the analysis of complex cyber threats, sharing insights and coordination of response efforts. From this perspective, this collaborative approach will help organizations to leverage the collective wisdom of their workforce for making better and more timely decisions related to cyberattack threats.

SDL provides a harmonious interaction between humans and machines in cybersecurity decision-making. Symbolic reasoning methods combined with deep models produces explainable predictions. That enables the cybersecurity professional to validate outputs from the models for possible biases or errors and to offer corrective feedback. This human-machine interaction builds in trust, transparency, and accountability into the decision-making process and enhances, in the end, the efficacy of cybersecurity operations.

SDL democratized cybersecurity expertise, and now any Information Technology (IT) administrator, business leader, or even a policymaker can contribute to decision-making processes. SDL expresses cybersecurity data and predictions in very transparent terms that are easy to understand, thus allowing non-experts to effectively contribute towards the management of cyber risks, threat detection, and incident response strategies. It is democratization that empowers an organization's resilience against cyberattacks by leveraging collective knowledge and expertise from all key stakeholders.

3.11.2 Democratization of Cyber Security

Ensuring that cyber security expertise and its tools are available to each and all—not only to experts—is one way forward to securing our digital world. SDL serves this purpose well by rationalizing decision-making systems for better interpretability and usability. These SDL models explain the reasoning behind their decisions, making it easier for non-experts to understand the basis of each recommendation. In that regard, SDL enables users with minimal knowledge of cybersecurity to make informed decisions for protecting themselves from cyber threats.

The integration of SDL in programs is a way of teaching students cyber security skills through practice and analysis of data to translate the outcome of these models in decision-making. Some of the SDL methods could help students address a cyber security issue in their profession by providing these students with relevant skills. Teamwork and creativity are allowed in the field of cybersecurity with open-source frameworks and tools by SDL. Advanced state-of-the-art technologies are made at everybody's disposal. Open-source SDL initiatives are constructed to make it easy for people to gain access to updated cybersecurity tools—anyone; newbies or experts alike from any background.

SDL empowers collaboration across functions for more informed decision-making in cyber security among IT professionals, business leaders, and policymakers. SDL would have diverse people from different backgrounds bring their own distinct perspectives to the table on any relevant cyber security discourse to ensure problems are effectively solved. In an integrated user-friendly interface, SDL models open access to advanced tools of cyber security, thereby opening cyber security to all—business or personal—without requiring specialist knowledge. In turn, a user-friendly interface would let the everyday person take control over online security by leveraging the power of SDL.

The SDL can use crowdsourcing to outsource threat identification tasks to the cybersecurity community at large, thus leveraging the power of the crowd. On the contrary, SDL offers an active role for individuals in threat intelligence efforts toward a jointly secure digital environment.

3.11.3 Establishing Trust with Users and Decision Support

The world of cybersecurity knows few issues as paramount as establishing trust with users, coupled with efficient decision support. This way, stakeholders can act assuredly on the insights that security systems provide. SDL can contribute much to this aim due to the fact that it allows for transparent and easy-to-comprehend models. Models from SDL provide interpretable explanations as to their predictions and recommendations, this enables the user to understand why the system decided in one way or another. Through the explainability of a system understandable by the user, SDL builds trust in users by showing exactly how it reaches its decisions and empowers user validation of system outputs (i.e., gives users the ability to evaluate the system's decisions for themselves).

SDL emphasizes transparency by giving users the chance to examine the inside workings of the model, specifically, users are able to observe how it takes any input data and returns any prediction. By making the inner mechanics of the model evident, SDL gains confidence by way of users who can independently evaluate the reliability and accuracy of outputs. In turn, validation and verification mechanisms can be incorporated in the SDL models for prediction reliability and robustness. The SDL further increases the ability to build trust by providing evidence of effectiveness and accuracy of the system through validation of the model's performance against known benchmarks or test scenarios provided to users.

The SDL will help teams make decisions uniformly by giving clear and understandable information with regard to cybersecurity data. It improves decision support, as SDL allows for the sharing of results among stakeholders, thus enabling them to collaborate and pool insights in light of what the system produces. It means SDL models can be put into easy-to-use interfaces and applications that see the information at hand clearly and intuitively visualized. SDL improves decision support by providing user-friendly ways to interact with the system, ensuring that users can easily access and understand the insights provided by the model.

It can implement emerging standards for explainable AI, ensuring that system outputs are transparent and easy to understand, trustworthy. Following guidelines and best practice for explainable AI will help the system build trust following a commitment to transparency, accountability, and user-centered design.

3.12 CONCLUSION

In summary, SDL is one historical footprint toward providing a comprehensive framework for cybersecurity decision-making. This new genre of ML marries advanced ML techniques with the clarity and transparency that symbolic reasoning brings. In this chapter, we have discussed various elements of SDL, such as its methodology, system architecture, implications, and applications in improving cybersecurity operations. The potential for SDL to be effectively applied toward critical challenges in cybersecurity decision-making is very encouraging. These include the demand for interpretable models, effective threat detection, and reduction of the equity gap across stakeholders. SDL models give explanations of their predictions in a very clear manner by integrating symbolic reasoning with deep learning architecture. This would help the cybersecurity professional to understand and validate, with supporting insights, what such systems generate for action.

SDL democratizes cybersecurity expertise to a larger section of users by bringing transparency and explanation to non-experts, hence making collaborative decision-making in the cybersecurity domain possible. In enhancing trust between users and decision support in cybersecurity, SDL is prominent as an indispensable tool. With interpretable explanations, transparent model operation, and validation processes, SDL also offers collaborative decision-making, ease of use through user-friendly interfaces, and standards for explainable AI.

If we are to push deeper, more research and development in SDL is required in order to overcome the rest of the challenges existing in this space, to unlock its full potential in cybersecurity. At this rate of innovation and refinement in SDL techniques, the state of the art in decision support for cybersecurity will have advanced, and the digital ecosystem will achieve a humanly resilient and more secure status for all. That is, SDL is not just about a technological step forward; rather, it is about changing how we make decisions about cybersecurity. Adopting SDL principles and methods will help with clarity, confidence, and trust as to how to navigate through each intricacy in the cyber landscape.

REFERENCES

[1] Muggleton, S. H., & De Raedt, L. (1994). Inductive Logic Programming: Theory and Methods. *The Journal of Logic Programming*, 19–20, 629–679.

[2] Marcus, G. (2018). Deep Learning: A Critical Appraisal. *arXiv preprint arXiv:1801.00631*.

[3] Yampolskiy, R. V. (2016). Turing Test as a Defining Feature of AI-Completeness. In B. Steunebrink, P. Wang, and B. Goertzel, Eds. *Artificial General Intelligence* (pp. 303–312). Springer, Cham.

[4] Anderson, K., Bowen, B. M., Coleman, R., Yung, M., & DeLong, R. (2016). Challenges and research directions in cybersecurity decision-support systems. In *Proceedings of the 49th Annual Hawaii International Conference on System Sciences*.

[5] Kwon, J., Lim, H., Shin, H., & Kim, J. (2017). Cyber Security Data Analysis: Challenges, Techniques, and Systems. *IEEE Access*, 5, 29006–29026.

[6] Ross, R. J., Hines, J. W., & Moore, A. P. (2017). Machine Learning and Cybersecurity. *IEEE Security & Privacy*, 15(4), 61–64.

[7] Papernot, N., McDaniel, P., Sinha, A., & Wellman, M. P. (2018). Deep Learning in Security: A Survey. *IEEE Security & Privacy*, 16(3), 38–49.

[8] Vasilomanolakis, E., Karuppayah, S., Mühlhäuser, M., & Fischer, M. (2015). Taxonomy and Survey of Collaborative Intrusion Detection. *ACM Computing Surveys*, 47, 1–33. [CrossRef]

[9] de Assis, M. V. O., Carvalho, L. F., Rodrigues, J. J. P.C., Lloret, J., & Proença, M. L., Jr. (2020). Near Real Time Security Systems Applied to SDN Environments in IoT Networks Using Convolutional Neural Networks. *Computers & Electrical Engineering*, 86 , 106738. [CrossRef]

[10] Rehman Javed, A., Jalil, Z., Atif Moqurrab, S., Abbas, S., & Liu, X. (2020). Ensemble Adaboost Classifier for Accurate and Fast Detection of Botnet Attacks in Connected Vehicles. *Transactions on Emerging Telecommunications Technologies*. [CrossRef]

[11] Manimurugan, S., Al-Mutairi, S., Aborokbah, M. M., Chilamkurti, N., Ganesan, S., & Patan, R. (2020). Effective Attack Detection in the Internet of Medical Things Smart Environment Using a Deep Belief Neural Network. *IEEE Access*, 8, 77396–77404. [CrossRef]

[12] Perez, M. G., Celdran, A. H., Ippoliti, F., Giardina, P. G., Bernini, G., Alaez, R. M., ChirivellaPerez, E., Clemente, F. J. G., Perez, G. M. & Kraja,

E., et al. (2017). Dynamic Reconfiguration in 5G Mobile Networks to Proactively Detect and Mitigate Botnets. *IEEE Internet Computing*, 21, 28–36. [CrossRef]

[13] Chung, Y. Y., & Wahid, N. (2012). A Hybrid Network Intrusion Detection System Using Simplified Swarm Optimization (SSO). *Applied Soft Computing*, 12, 3014–3022. [CrossRef]

[14] Aburomman, A. A., & Reaz, M. B. I. (2016). Review of IDS Development Methods in Machine Learning. *International Journal of Electrical and Computer Engineering* (*IJECE*), 6, 2432–2436. [CrossRef]

[15] Bijalwan, A. (2020). Botnet Forensic Analysis Using Machine Learning. *Security and Communication Networks*, 2020, 9302318. [CrossRef]

[16] Alothman, Z., Alkasassbeh, M., & Al-Haj Baddar, S. (2020). An Efficient Approach to Detect IoT Botnet Attacks Using Machine Learning. *Journal of High Speed Networks*, 26, 241–254. [CrossRef]

[17] Ford, V., & Siraj, A. (2014). Applications of machine learning in cyber security In *Proceedings of the 27th International Conference on Computer Applications in Industry and Engineering.*

[18] Jiang, H., Nagra, J. & Ahammad, P. (2016). Sok: Applying Machine Learning in Security-a Survey. *arXiv preprint arXiv:1611.03186.*

[19] Ribeiro, J., Saghezchi, F. B., Mantas, G., Rodriguez, J., Shepherd, S. J. & Abd-Al Hameed, R. A. (2020). An Autonomous Host-Based Intrusion Detection System for Android Mobile Devices. *Mobile Networks and Applications*, 25(1), 164–172.

[20] Hodo, E., Bellekens, X., Hamilton, A., Dubouilh, P. L., Iorkyase, E., Tachtatzis, C., & Atkinson, R. (2016). Threat analysis of IoT networks using artificial neural network intrusion detection system. In *Procroceeding 2016 International Symposium on Networks Computers and Communications (ISNCC)*, IEEE, pp 1–6.

Chapter 4

Can Neural Networks Dream?

Applying Crick and Mitchison's Theory of Reorganization to Multi-Layer Perceptrons for Reduction of Catastrophic Interference

Ashwini Dalvi, Aatmaj Kaustubh Mhatre, and Radha Shirbhate

4.1 INTRODUCTION

Dreams are one of the most complex processes that occur in our brain. There are many theories explaining the reason behind the occurrence of dreams. One such theory that explains why dreams occur is the Crick and Mitchison theory. This theory tries to explain the underlying mechanisms occurring in the brain during dreams. According to this theory, a mechanism known as reverse learning occurs in our brains while we dream. This explains both the unconventional nature of dreams and the inability to recall them accurately. According to the theory [8], dreams that occur in REM sleep[1] are an attempt at removing unnecessary information (parasitic modes) from the brain [28]. This theory postulates the process of removing unwanted modes—reverse learning[2]—as one of the underlying mechanisms that cause dreams in REM sleep [7, 21]. According to Crick and Mitchison, the goal of the reverse learning mechanism is to prevent the brain from getting overloaded with too many memories, as they say, "We dream in order to forget" [9, 24, 29]. Counterintuitive to the common intuition that REM sleep strengthens the connections between brain neurons, the reverse learning mechanism actually aims to weaken the connections. During this mechanism, random signals are given to the brain. These random signals are known as ponto-geniculo-occipital (PGO) waves [19, 30, 42][3]. The connections of neurons of the brain that respond to these waves are dampened down. This is, in essence, the opposite of the learning process, where the connections between neurons which fire together are strengthened. The theory also postulates that this mechanism serves to reduce network "obsessions"[4]. If the Crick and Mitchison theory of dreams is so critical for biological brains, can it be applied in artificial neural networks (which, for reference, represents the full form of ANNs, but will be referred to henceforth as artificial neural networks throughout this article)? Can artificial brains dream too? The objective of this study is to apply Crick and Mitchison's theory of reorganization in artificial brains by making neural networks dream and observing its positive effects if any.

DOI: 10.1201/9781003501220-4

According to this theory, the overload of the brain is reduced by reverse learning. Similar to the overloading of natural brains, artificial neural networks also get overloaded. One of the areas of application where the overloading of neural networks causes problems is lifelong learning. The ability of neural networks to continuously learn and update their decisions based on new data is called lifelong learning [35]. A neural network capable of lifelong learning can learn even when data of different orientations is provided sequentially to the model. When new types of data are fed into the model, it should be able to learn the new data, as well as remember the old data without requiring retraining from scratch. Applications of artificial intelligence (AI) in areas like natural language processing or self-driving vehicles where the nature of the data changes rapidly require lifelong learning. The same models are required to adapt to dynamic environments without exhaustive re-training. However lifelong learning is a challenge for artificial neural networks. Artificial neural networks currently in practice cannot learn continuously for a long period of time. These networks tend to "forget" at a very rapid rate.

The reason behind this rapid forgetting is known as catastrophic interference or catastrophic forgetting. Catastrophic interference is a phenomenon that prevents neural networks from lifelong sequential learning [32, 40]. This effect arises when the new learning data interferes with old network connections and causes the network to drastically forget what it had learnt earlier. For example, if a model is first trained on data A, and then on data B, the backpropagation algorithm causes the weights of A to be erased. This is because while training for data B, the model focuses on minimizing the error rate of B only, without concern for any of the previous data. This erasing occurs at a rapid rate, hence the term catastrophic. When the old data is erased, the model requires retraining on the old data, leading to a huge loss of performance.

While continual learning remains a challenge for neural networks, biological brains are capable of continually acquiring knowledge without forgetting previous knowledge [33]. They can easily adapt to new environments without needing to retrain. On learning new facts, biological brains are capable of learning them at the same time remembering the old facts. Forgetting, if any, is very small in size and is not as drastic as that of artificial brains. One of the factors that prevent dreaming in biological brains is REM sleep. What if REM sleep is introduced in artificial neural networks? Can dreaming be a solution to this problem in artificial brains as well? Recent studies have shown how sleep can be effective against catastrophic forgetting in spiking neural networks [16, 38. Our study takes inspiration from this idea. Application of Crick and Mitchison's dream theory (reverse learning) in artificial networks does reduce the rate of catastrophic forgetting.

In our research, the reverse learning mechanism is applied to multilayer perceptions (MLP). The model is used to train the MNIST (Modified

National Institute of Standards and Technology) handwritten digit database for optical character recognition [41]. To simulate sequential learning, the entire training dataset is split into two parts. The model is first trained sequentially first on the first part and then on the second. The rate of catastrophic forgetting is measured across a number of samples. To identify if dream sleep has any effect on catastrophic forgetting, the same process is repeated over. But this time, after the model is trained on the first part, dream sleep (reverse learning) is introduced. Rapidly alternating sessions of a few samples of learning are followed by many noise signals (sleep). Various cases are discussed in the results and discussions section. Performance of the model with sleep and without sleep are compared by measuring accuracy drop across sample size. Accordingly, this result concludes that dreaming does help reduce the rate of catastrophic forgetting.

The main contributions of the study are listed below:

1. By applying the Crick and Mitchison reorganization theory, dream sleep—reverse learning algorithm as an effective method of reducing catastrophic forgetting is presented. The reverse learning mechanism for artificial neural networks is discussed in detail. Its implementation is compatible with the current training methods which have been provided.
2. Reverse learning has been applied to the MNIST database. The corresponding reduction in catastrophic forgetting was measured using accuracy graphs. The results were analyzed.
3. Various parameters for the reverse learning mechanism have been discussed.

4.2 LITERATURE SURVEY

Catastrophic interference is the tendency of neural networks to forget previously learnt data at sudden and steep rates. Causes and prevention of catastrophic interference have been extensively studied after its discovery by McCloskey and Cohen in 1989 [32]. Some of the significant studies include works by French [11] and Kemkar [26]. Methods of prevention of catastrophic interference were proposed by French [12] and Kirkpatrick [27]. A recent study by Parisi on lifelong learning has also attracted attention [35]. Goodfellow [18] investigated the extent to which the catastrophic forgetting problem occurred in neural networks for established and new algorithms. Andrew Robins [36] tried to overcome this problem using the rehearsal methods, which use previously learned data again for reducing catastrophic interference. Research by Tyler Hayes [22] proposes a brain-inspired approach "REMIND" which replays compressed memories. According to the authors, the REMIND's replay is analogous to the replay that occurs in the brain during waking hours, and it would be interesting to explore how to effectively create a variant that utilizes sleep/wake cycles (REM sleep) for replay. Moreover, they state that this could be particularly advantageous for an agent who is mostly engaged in online learning during specific hours and offline consolidation during other hours. This REM sleep replay is something similar to what is proposed in our study.

Mandivarapu [31] provides a new method for training data—Self-Nets which claims to incorporate new data with little retraining, and minimal forgetting, without storing prior training data. However, Self-Nets are novel concepts and thus do not provide compatibility with old techniques. Our study uses the backpropagation algorithm to implement the dream sleep algorithm, hence not requiring the implementation of any new methods in pre-existing implementations.

Furthermore, recent studies have mapped sleep with catastrophic interference [2, 4]. Tadros developed a sleep algorithm for spiking neural networks for the reduction of catastrophic interference [38]. Experiments related to random noise images using pseudo-rehearsal on SNN (spiking neural network) were conducted by Antonov [5]. Gonzalez tried to simulate sleep in SNN for improvement of continual learning by combining consolidation of new memory traces with re-consolidation of old memory traces [17]. They tried to use slow-wave sleeps (SWS) to reduce catastrophic interference. Golden simulated sleep in a very similar manner as our research—by providing random noise to SNN neurons but reinforcing old knowledge without using the old data [15]. Watkins used sinusoidally modulated noise as an analogue to sleep. A few studies on neural networks and dreams have also been presented. Fachechi tried to simulate dreams on Hopfield nets for increasing storage capacity [10]. However, none of these studies presents reverse learning as an effective medium against catastrophic forgetting. The majority of these studies have used some kind of "reinforcing" to improve the model by reusing the old data. However, these methods will not be useful in the condition that the old data cannot be stored in any form, compressed or otherwise (i.e., these approaches require some version of the original training data to prevent forgetting. If the old data cannot be saved or reused in any way, the methods become ineffective). On the other hand, the dream sleep algorithm presented in our study does not utilize old training data in any format.

The reverse learning mechanism was first proposed by Crick and Mitchison [7, 8, 9]. They simulated the mechanism on artificial neural networks for research on biological neural networks. Studies like these have used artificial neural networks to prove concepts of biological neural networks [3]. Christos used Hopfield neural networks to prove that Crick and Mitchison's theory reduces overloading [6]. All these studies were applications of artificial neural networks as a technique for simulating biological brain systems. However, a study which applies their theory to artificial neural networks or aspects of AI or machine learning (ML) could not be found. In this study, the application of their theory has been done to try to reduce catastrophic interference.

4.3 DREAMING NETWORK

4.3.1 Problem Statement

Consider two separate categories, explicitly, data A and data B. A data set containing data belonging to either category is split into corresponding two datasets A' and B'. Example A has odd digits while B has even digits. Then dataset A' contains samples of images of odd digits while dataset B' contains samples of even digits. A simple multi-layer perceptron model is trained on the data. The aim of the model is to classify an image that belongs to either of the following categories is formed by $A \cup B$.

The model is first provided with dataset A and then with dataset B. The assumption here is that the worst-case scenario is present. The following assumptions are made:

1. The model cannot store any training data
2. Sets A and B have no element in common $A \cap B = \emptyset$
3. Categories of data in set A and set B have no element in common, that is sets A' and B' have no element in common $A' \cap B' = \emptyset$
4. Data in set A is provided sequentially followed by data in set B

4.3.1.1 Catastrophic Forgetting

Since the storage of the data is not permitted, the training process involves sequentially training the model on dataset A, followed by dataset B. However, such a sequential form of learning on two datasets (A followed by B) results in catastrophic forgetting. When the model is trained on dataset A. It forms associations related to categories of data in A. Thereafter, when it is trained on dataset B, it forms associations related to categories of dataset B. While forming these associations, it erases the associations formed earlier. Hence it forgets the knowledge gained when trained on A. This can be inferred from the research by [32]. In their case, they have used the addition of one as A and the addition of two as B (i.e., they designated the addition of one as dataset A and the addition of two as dataset B).

4.3.2 Reverse Learning Dream Sleep algorithm

The reverse learning mechanism makes the model "forget" connections. The forward propagation and the backpropagation algorithms, that form the learning algorithm give invalid inputs and process the network. This increases the neuron connections on valid firings and reduces the connections on invalid firings. The reverse learning mechanism acts just the opposite to the forward propagation method. It provides the model with random noise as input. Then it processes the model by reducing any connections that fire together. The reverse learning mechanism is based on the essence proposed by the Crick and Mitchison theory [7]. The random noise given to the network is an analogue to the PGO (which, as a reminder to the read, stands for, ponto-geniculo-occipital) waves that occur in our brain.

According to them, the process of reverse learning is, the dampening of physical neuron connections in the brain by PGO waves. During sleep, PGO waves acting as a random noise are fed into the biological neural network. Then the connection strengths between the neurons that fire in response to this random noise are reduced (i.e., neurons that fire in response to random noise have their connections weakened to minimize the impact of irrelevant activity).

A similar reverse learning process can be implemented for artificial neural networks. Firstly, provide random values as input to input neurons, with the range of the values depending on the model. Then forward propagate the values in the network through all layers. Reduce the weights of the neurons that show a high value. These steps can be easily translated to use methods that are already in practice, that is the standard forward learning algorithms—forward propagation and backward propagation. This is because backpropagation has the same effect of reducing the weights—if the expected values of all output neurons are made zero. This can be achieved by providing a label that is not present in either of the sets to the model. The methodology for reverse learning is explained Algorithm 4.1. How the entire dream sleep algorithm works is explained in Algorithm 4.3.

In reverse learning, unnecessary connections are reduced but not completely erased. We can simulate the reverse learning mechanism on artificial neural networks using a set of simple procedures, as proposed in the re-organization theory [9] and modified for application in multi-layer perceptrons.

Step 1. Provide random noise as input to the model.
Step 2. Set the label to k where k is a label which is not present in the training data and corresponds to the expected activation of all output neurons to 0. (Example in this case label can be '10' or 'A')
Step 3. Forward propagate the model with the random values.
Step 4. Set the learning rate to sleep learning rate. Backward propagate the model.

Here steps 2 and 4 have the same effect as adjusting the synapses between neurons to reduce the association slightly. This is because by setting the expected answer (label) to something the model cannot predict, (Example setting label '10' or 'A' when a model can predict only numbers from 0 to 9) the expected activation values of the output layer are set to zero. When the model predicts any of valid predictions for the noise, that value must be reduced. Example if the model predicts eight, it has an obsession for value '8' or neurons associated with value '8' are overloaded. The connections of all such neurons must be reduced. This reduction is automatically handled by the backpropagation algorithm, as the expected values are set to zero.

Algorithm 4.3 summarizes the reverse learning procedure.

Algorithm 4.1 Reverse Learning

Input: Model trained on a Dataset D. Model is designed for total possible labels (classes) set S with total dimensionality *d*. Model accepts numerical input values in range (a,b). The model is a classifier model that predicts the probability of the each class in S.

Output: Model with single iteration of reverse learning performed on it. (Model with 1 dream cycle introduced on it.)

Parameters: Rate of forgetting α

Steps:

1. Generate a d dimensional datapoint with random values in the range (a,b)
2. Assign a label l to the datapoint such that $l \notin S$
3. Feed forward the datapoint in the model
4. Backpropagate the model with learning rate = α

The below section represents the mathematical representation of this algorithm.

4.3.3 Mathematical Representation of Reverse Learning

The reverse learning mechanism can be applied to an artificial neural network. The reverse learning mechanism is defined using mathematical notations for a multi-layer perceptron that trains using the forward and backward approaches.

4.3.3.1 Definitions

$X_{i,j}$ = Activation of ith neuron in layer j.

We assume that the activation has a range from -1 to 1, except for input and output neurons, which are $X_{i,j} \in (-1,1)(\forall l > j > 0)$

$W_{i,j,n,m}$ = Weight between the ith neuron in layer j an
α = Rate of forgetting (analogous to learning rate)
l = Number of layers
n_j = Number of neurons in layer j
$b_{i,j}$ = Bias corresponding to i^{th} neuron in layer j
$\delta_{i,j}$ = Error corresponding to i^{th} neuron in layer j

Let h represent a strictly monotonically increasing rectification function in the range 0,1.

4.3.3.2 *Forward Pass*

The forward pass works as follows, provide random values as input to the input neurons. The range of the values depends on the model. Forward propagate the values in the network through all layers.

For a general case of a network operating in the range 0–1, the random inputs can be represented as:

$X_{i,0}$ = Random input between 0 and 1

Feeding these random numbers into the network and applying forward propagation (weighted sum of neuron outputs) the following expression is obtained:

$$X_{i,j} = h\left(\sum\nolimits_{k=0}^{n_j}\left(W_{i,j,k,l-1} \cdot X_{k,l-1}\right) + b_{i,j}\right) \tag{1}$$

4.3.3.3 *Backward Passs*

Reduce the connections that fire to the random noise. This can be done by using the delta learning rule, with no output as the expected output. That is by implementing standard backpropagation procedure by setting the expected values of all output neurons to zero.

This results in the following expression for the updated weights:

$$W_{i,j,k,j-1}^{+=W_{i,j,k,j-1}+\alpha \cdot \delta_{i,j} \cdot X_{k,j-1}} \tag{2}$$

Where the error function is given by:

$$\delta_{i,j} = X_{i,j}\left(1 - X_{i,j}\right)\sum\nolimits_{k=0}^{n_j}\left(\delta_{k,j+1} \cdot W_{i,j,k,j+1}\right) \tag{3}$$

$$\delta_{i,l} = -\left(X_{i,l}\right)^2\left(1 - X_{i,l}\right) \tag{4}$$

4.3.4 Training using Dream sleep

We are considering the case of a continuous learning model, we have the initial dataset 'A' on which the model is trained. After the model is trained, we have to retrain it on data points that appear as time progresses. Intermixing a few parts of A and many parts of B can be done. However, in this study, we are considering a worst case scenario where the model does not return to dataset 'A' after it is trained, and is trained only by 'B' data. We are also considering the worst case where A and B datasets are mutually exhaustive, that is, the type of data that occurs in B does not appear in A and data in A does not appear in B.

By introducing dream cycles while training the dataset B, the rate of catastrophic forgetting can be reduced. The methodology is explained in the Flowchart 4.1. The detailed algorithm is provided in Algorithm 4.3. The process introduces k samples of sleep after every n sample of training of dataset B.

If the model had been trained without dream cycles, then the algorithm would have been a traditional training algorithm, as explained in Algorithm 4.2.

The dream sleep algorithm works differently. Part A is trained first as traditional method suggests[5]. It is, however, followed by rapid alternating sessions of B and sleep. The sequential order of A-B is followed, meaning that, once 'A' is trained, then the data related to/part of A is completely excluded from the model.[6]

The training algorithm with the incorporation of sleep can be expressed as follows:

Step 1. Train dataset A.
Step 2. Train n elements from dataset B.
Step 3. Run sleep algorithm k times.
Step 4. Repeat steps 2 and 3 until dataset B is exhausted.

The training algorithm can be summarized in Algorithm 4.3.

4.3.4.1 *Train to Sleep Ratio*

After n successive training of datapoints, k iterations of sleep are performed. The ratio n:k can be identified as one parameter which affects the performance of the model. The value of the batch-size also plays a significant role on the performance, especially when the value of n is comparable with the batch-size. On an experimental basis, it has been found that training two samples, followed by sleep for 20 cycles is an optimal solution, however, 20 samples followed by 20 sleep cycles is more efficient. Reducing the value of n and increasing the value of k increases the amount of processing required.[7]

Algorithm 4.2 Training without dream sleep

Input: Datasets A, Set of sequentially available datapoints B, classifier model M. Dataset A is available only in the initial stage. B is available as the points emerge. Neither A nor B can be stored. A and B have mutually exhaustive classes.

Output: Trained model that can classify only B.

Parameters: Model architecture

Steps:
1. Train on Dataset A
2. Train on Dataset B
3. End

Algorithm 3 Training with dream sleep

Input: Datasets A, Set of sequentially available datapoints B, classifier model M. Dataset A is available only in the initial stage. B is available as the points emerge. Neither A nor B can be stored. A and B have mutually exhaustive classes.

Output: Trained model that can classify both A and B.

Parameters: Number of training points to be considered before sleep: n
Number of sleep cycles: k
Rate of forgetting: α

Steps:
1. Train on A
2. **While** samples of B remaining:
   a. Train n samples of B
   b. **For** k times:
      Perform Reverse Learning on M with rate α
3. End

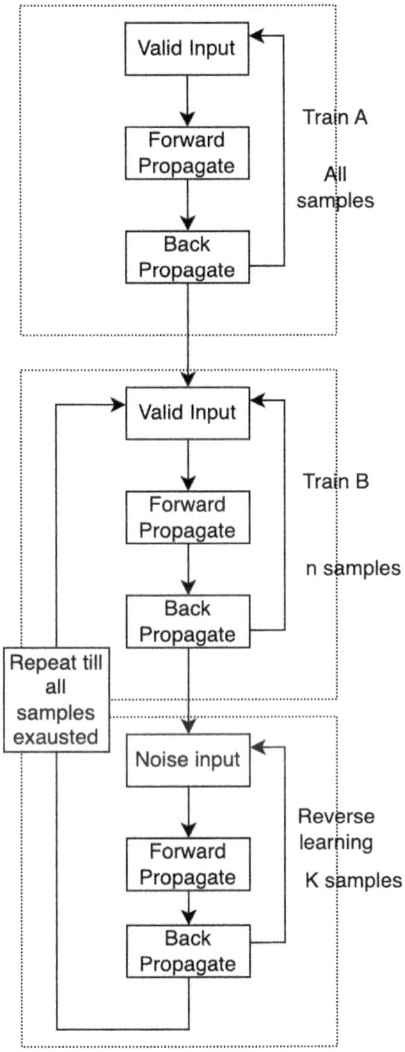

Figure 4.1 Dream Sleep Reverse Learning Methodology Flowchart.

4.3.5 Relation with Regularization

Considering the principle behind the dream theory, which is that dreaming prevents overstraining or overburdening of the brain, the dream sleep algorithm can be interpreted as a form of regularization. Regularization is a method employed in ML and deep learning (DL) to avoid overfitting and enhance the model's ability to generalize. It improves the performance of the model by reducing the variance in the model. It makes the model simpler. The dream cycles that are introduced have a similar effect of making the model simpler by removing the unwanted connections in the neurons. Due to the regularization effect of the dream cycles, the overfitting of the model on the available training data that is available is reduced, so that it can accommodate new training data that is made available at a later stage.

4.3.5.1 Weight Decay

In weight decay regularization techniques, weights are reduced and taken towards zero. A neural network featuring smaller weight matrices results in more straightforward models that significantly minimize the issue of overfitting [25].

Weight decay works by adding a penalty term to the cost function.

$$\delta'_{i,j} = \delta_{i,j} + \lambda * \omega(i,j)$$

As lambda increases, weights decrease. Where omega is a penalty term that is a function of the trainable model parameters [34,39]. The two most common techniques of weight decay are L1 and L2 regularization. Both L1 and L2 regularization work towards reducing the weights of the neural network. L1 regularization (LASSO) reduces the weights by changing the cost function to force the weights to zero. L2 regularization, also referred to as ridge regression, is a technique that imposes a penalty on model parameters by adding a term to the loss function. This penalty term causes the model's weights to be "decayed" or pulled towards zero, effectively shrinking their magnitude. This decay is only partial, allowing for some non-zero values that still contribute to the model's predictions.

However, the dream sleep algorithm does not work in the same manner. Rather than reducing the absolute value of all weights, it reduces only those weights that respond to the stimulus of random input. Weights in the hidden layer may increase or decrease depending on the situation. This means that the importance of the weights that are responsible for producing a valid output when the input itself is invalid is reduced drastically. While the importance of other weights will not change much.

For the last layer, we can prove that all the weights will be taken towards a negative value.

Since:

$$W^{+}_{i,j,k,l-1} = W_{i,j,k,l-1} + \alpha \cdot \delta_{i,l} \cdot X_{k,l-1} \tag{5}$$

$$W^{+}_{i,j,k,l-1} = W_{i,j,k,l-1} + \alpha \cdot -\left(X_{i,l}\right)^{2}\left(1 - X_{i,l}\right) \cdot X_{k,l-1} \tag{6}$$

Since $X_{i,j}$ is in range (0.1), all weights will reduce. For hidden layers, the value of the weights depends on the weights of the next layer, thus depending on their values they might increase or decrease.

While training, the neural networks work on the principle—"neurons that fire together wire together". During dream cycles, the principle is that "connections between neurons that fire together are reduced" This is different from weight decay regularization where the absolute value of the weights is reduced and taken to zero.

4.4 EXPERIMENTAL SETUP

In this section, the dreaming methodology has been applied on the MNIST digit database. Detailed accounts of the implementation have been given.

4.4.1 Dataset

MNIST handwritten dataset [41] was used for training the data. This data set consists of handwritten digits 0–9 in grey-scale 28x28 pixel image format. 50000 samples of training data and 9500 different samples of testing data were used for this study. This is the standard MNIST database [41].

4.4.2 Model Architecture

The target accuracy we aim for in the architecture is 85–90%. Although various methods of increasing the accuracy beyond 95% are in practice, some of them including complex networks like CNN (Convolutional Neural Network) [14, 20]. The goal of this study is not to increase the overall accuracy of the systems, but to apply the dream theory to neural networks in order investigate its' effect on catastrophic forgetting. One of the main reasons for the decreased accuracy thought the study is due to the fact that every image is trained only once (for one epoch only) to simulate dynamic environments, especially those in which data cannot be stored.[8,9]

4.4.2.1 Neural Network

The supervised neural network used for training the data consists of 784 input neurons, 10 output neurons. Every output neuron corresponds to digit from 0–9. The network consists of two hidden layers of 16 neurons each. Stochastic gradient descent is performed with a batch-size of four samples. Weight and bias learning rate were set to one. The weights were initialized with small random numbers. For simulation of dynamic environments, every image was trained for one epoch only. Sigmoid activation function (The sigmoid activation function maps any input to a value between 0 and

1, creating an "S"-shaped curve. It is used for binary classification) was used. More details of the model and algorithms used can be found at the digital [37] sources and [23].

The model is implemented in Java by using ANN4j package [1].

4.5 RESULTS AND DISCUSSIONS

This section will include the experimental results of the theory presented. The dream sleep algorithm (Algorithm 4.3) was applied on the above described neural network and datasets. The results were then compared with the without dream algorithm (Algorithm 4.2). The results of various experiments performed on hyperparameters, and network configurations of the proposed dream sleep algorithm will be explained. The reduction in the rate of catastrophic interference will be presented in a visual manner. This section justifies how incorporating dream cycles into the training process helps the catastrophic interference.

4.5.1 Performance of Dream Sleep Algorithm

The following results can be seen through the experiments that were conducted.

4.5.1.1 Increase in Accuracy

1. The introduction of dream sleep cycles improves overall accuracy.
2. In both cases (with and without sleep), accuracy declines as more samples of class B are trained. However, accuracy without sleep is lower than accuracy with sleep.

4.5.1.2 No Effect on Recall of Class B

1. Dream cycles have no effect on the recall of class B, which means that the model does get trained on class B.

4.5.1.3 Decrease in Catastrophic interference

1. Recall of class A with sleep is with greater than recall without sleep, which means that the rate of forgetting is reduced.
2. The overall accuracy also proves that the model forgets at a slower rate when dream cycles are introduced.

4.5.2 Results for a Single Digit

Consider the case where training for nine digits was performed, and after which one digit was trained. The results for few digits are shown as samples in this section. One thing to note here is that the performances of the model across digits varies with respect to its characteristics in the dataset. These characteristics include the occurrence of every digit.

Number of samples of each digit are from zero to maximum possible training samples at intervals of 100 samples of B.

4.5.2.1 For Digit One

We train all digits except one first, and then only after all the other digits are trained do we train digit one.

A = {0,2,3,4,5,6,7,8,9}

B = {1}

Accuracy graphs for different configurations of sleep ratio are given in Figure 4.2. The x axis represents the number of samples from 0% to 100% while the y axis represents the accuracy of the model in percentages. Various train to noise ratios are color coded. Example 2:20 represents two samples followed by 20 noise images (dream sleep) colored blue. It can be seen that Catastrophic interference is reduced for different configurations of dream sleep. The red curve represents the performance without sleep. We see that the accuracy without sleep is 25% while with sleep it greater than is 50%. The results at 100% training are summarized in Table 4.1.

If the model is made to sleep in bursts, then the increment in accuracy at each bust can be clearly seen. Figure 4.3 shows one such result. After every 2000 samples, 2000 cycles of dream sleep were introduced. The effect at each spike is clearly visible. This further proves that the accuracy is increased by the dream sleep cycles.

Figure 4.4 shows a detailed version of model performance for the 2:20 configuration. The individual recall graphs for two digits, one and four are shown. There is no effect on the recall of class B (digit 1) while recall of class A (digit 4) is higher. At 100% training, recall of four without sleep is 10.72% while with sleep is 22.96%.

Table 4.1 Improvement in Accuracy at Various Training Levels and Dream to Sleep Ratios

Training Level	*Without Sleep*	*With Sleep*	*With Sleep*	*With Sleep*
		2:20	*20:20*	*200:200*
10%	41.23157895	80.08421053	55.13684211	55.31578947
25%	32.72631579	66.14736842	50.31578947	57.68421053
50%	26.74736842	55.50526316	45.15789474	54.67368421
100%	24.18947368	45.07368421	41.18947368	49.53684211

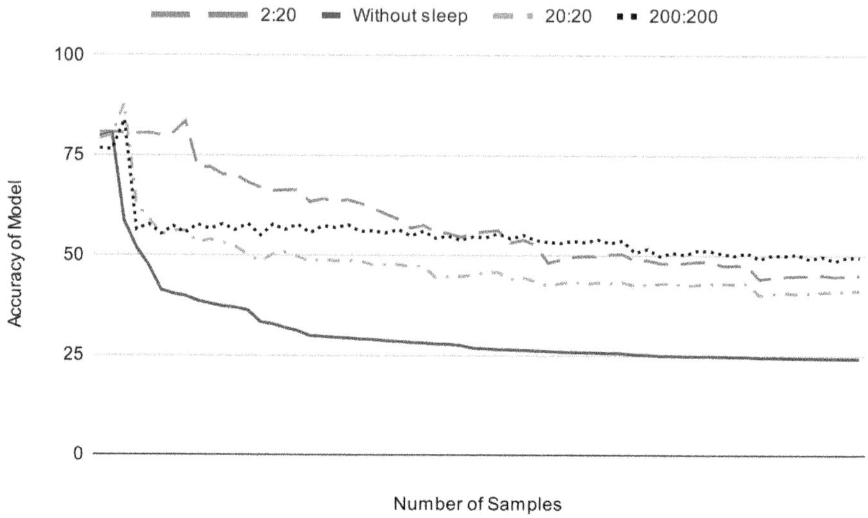

Figure 4.2 Accuracy vs number of samples for various train to noise ratios.

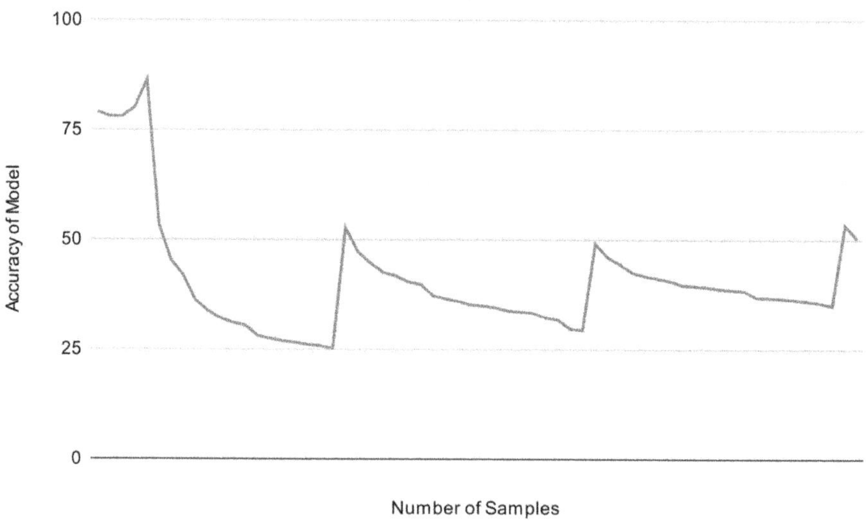

Figure 4.3 Accuracy vs Number of Samples.

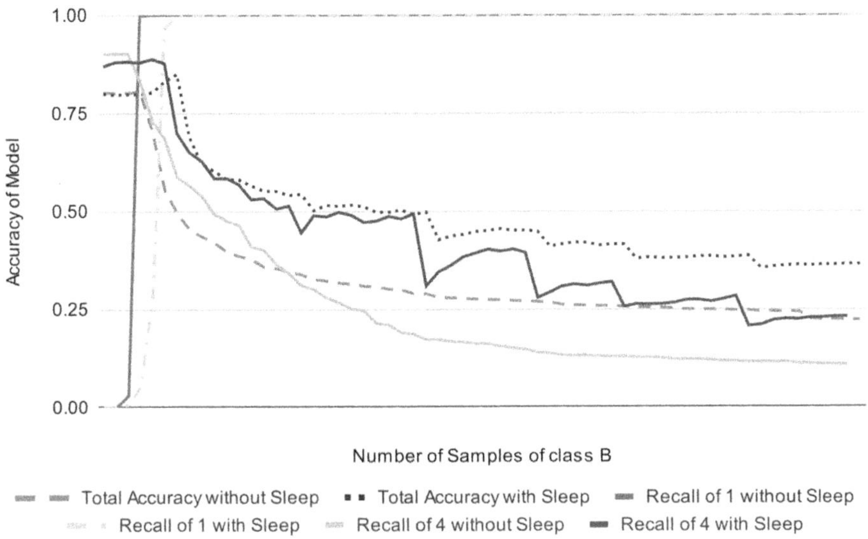

Figure 4.4 Accuracy and Recall vs Number of Samples for Various Configurations.

Figure 4.5 shows the model performance for a different configuration. This is a denser model with three hidden layers of 32, 16, 16 neurons each. This further shows improvement in accuracy. The red curve shows the accuracy with sleep, while the blue curve shows without sleep. The fact that the curve indicating results with sleep is higher than without sleep shows how the rate of the catastrophic interference is reduced. At 100% training, the accuracy without sleep is 22.23157895 while with sleep is 44.38947368. This is a 2x improvement in the accuracy.

4.5.2.2 For Digit Zero

One thing to note is that the improvement depends on the nature of the dataset and the configuration of the model. Hence, there are varying results in various models and datasets. However, one general observation is that towards the middle of the graph, the accuracy with sleep is greater than without sleep. Hence the results obtained for digit zero are different than that of digit one.

Graph for digit zero can be seen in Figure 4.6.

A = {1,2,3,4,5,6,7,8,9}

B = {0}

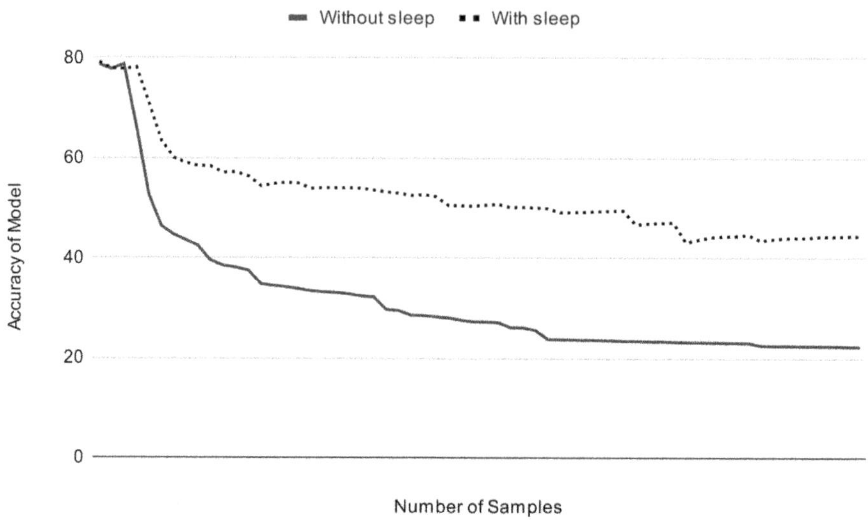

Figure 4.5 Model Performance for Digit One in a Different Configuration.

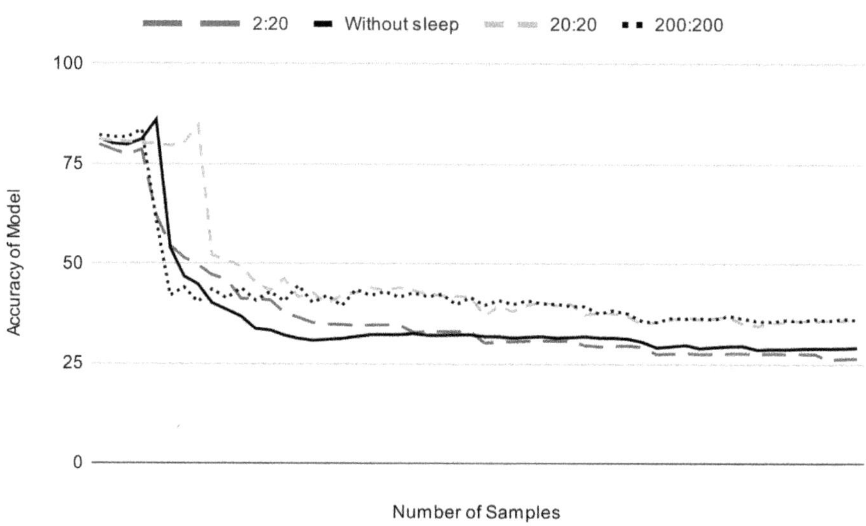

Figure 4.6 Accuracy vs Number of Samples for Digit Zero show a 10% Improvement in Accuracy.

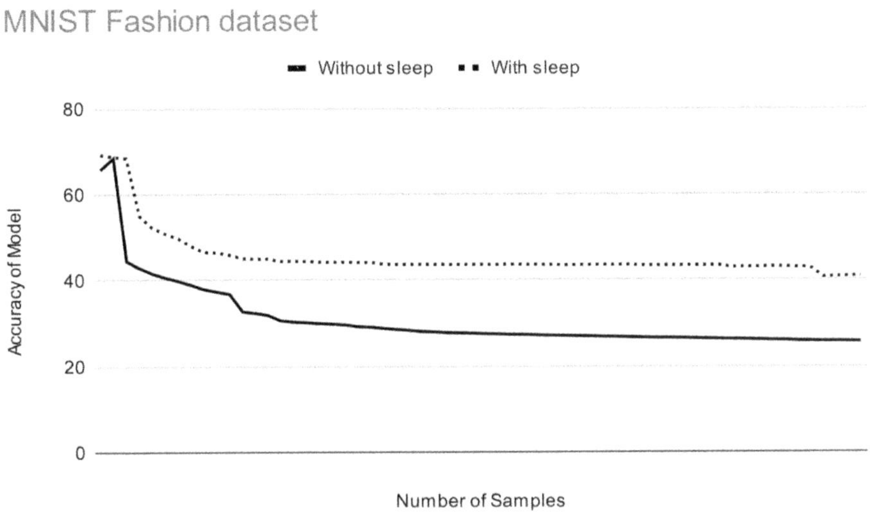

Figure 4.7 Model Performance for on MNIST Fashion Dataset.

We can see for some configurations, the accuracy increase by about 10% (1.4x improvement) is obtained by introducing dream sleep cycles.

4.5.2.3 Fashion Dataset

Figure 4.7 shows the reduction in catastrophic interference on the Fashion MNIST dataset. This dataset contains images of fashion products from 10 categories in standard MNIST format [44]. All other parameters of the model remain the same.

A = {T-shirt/top, Pullover, Dress, Coat, Sandal, Shirt, Sneaker, Bag, Ankle boot}

B = {Trouser}

Accuracy is increased from 25.63% to 40.96%.

4.5.3 Performance for Two Digits

The effects of catastrophic forgetting when set B contains two digits are shown for two combinations.

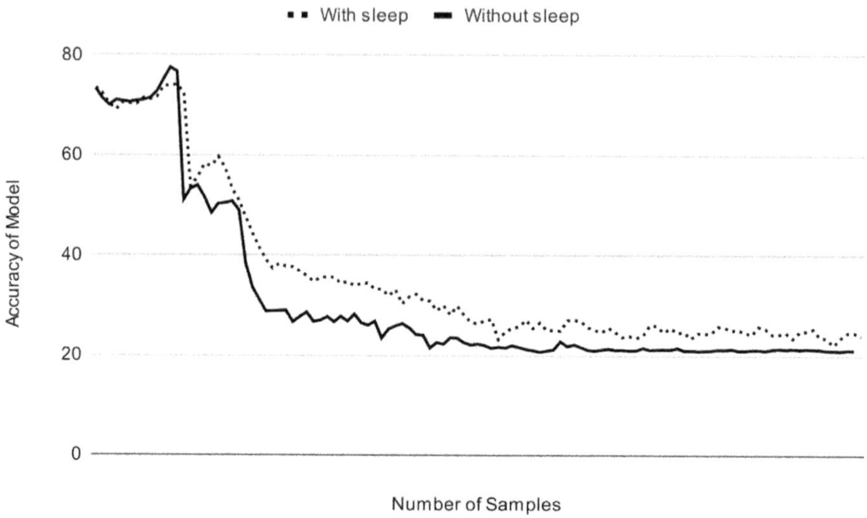

Figure 4.8 Model Performance for Two Simultaneous Training of Digit One and Five.

4.5.3.1 *One and Five*

Figure 4.8 shows improvement of the overall accuracy. In the middle section of the graph, accuracy is increased, although it tapers down towards the end. This means that the rate of catastrophic interference is slightly lower with sleep than without sleep.

A = {0,2,3,4,6,7,8,9}

B = {1,5}

4.5.3.2 *Zero and Five*

Figure 4.9 shows the results The accuracy is always greater with sleep than without sleep. At 100% training, accuracy is improved from 19.37% to 26.34%.

A = {1,2,3,4,6,7,8,9}

B = {0,5}

4.5.4 Sequential Learning for Three Sets

For further insight into how sleep performs for sequential learning, the dataset is divided into three sets instead of two. Training of subset A is

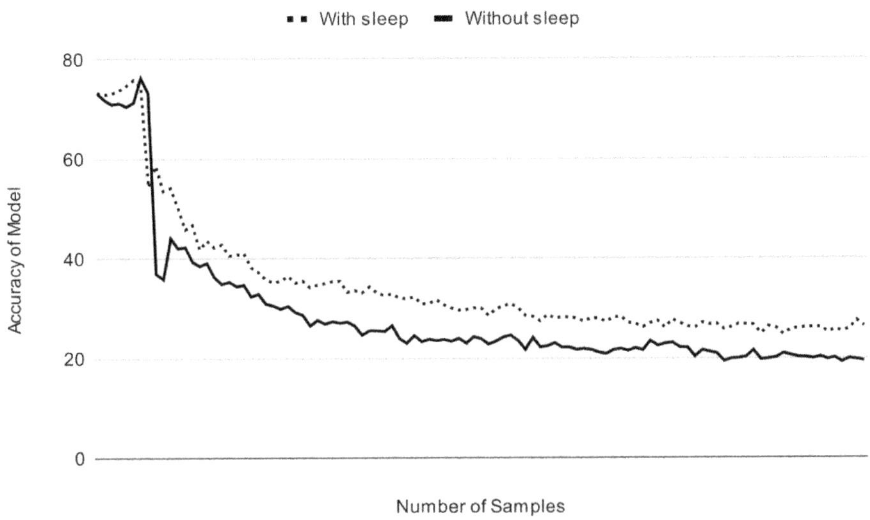

Figure 4.9 Model Performance for Two Simultaneous Training of Digit Zero and Five.

followed by subset B followed by the last subset C. No overlapping of data is present.

A = {0,2,3,4,6,7,8,9}

B = {1}

C = {5}

Three conditions are evaluated. First, no sleep cycles are introduced. Then sleep cycles are introduced between both B and C. The third case is where the sleep cycle is introduced only in the middle subset B. From the graph obtained in Figure. 4.10, it is visible (or, quantifiable) how dream sleep increases performance.

4.5.5 Comparison of Models with and without Sleep

Models with sleep show up to 2x accuracy measurement. This shows that the rate of catastrophic interference is significantly reduced. The performance of the models with and without sleep is summarized in Table 4.2.

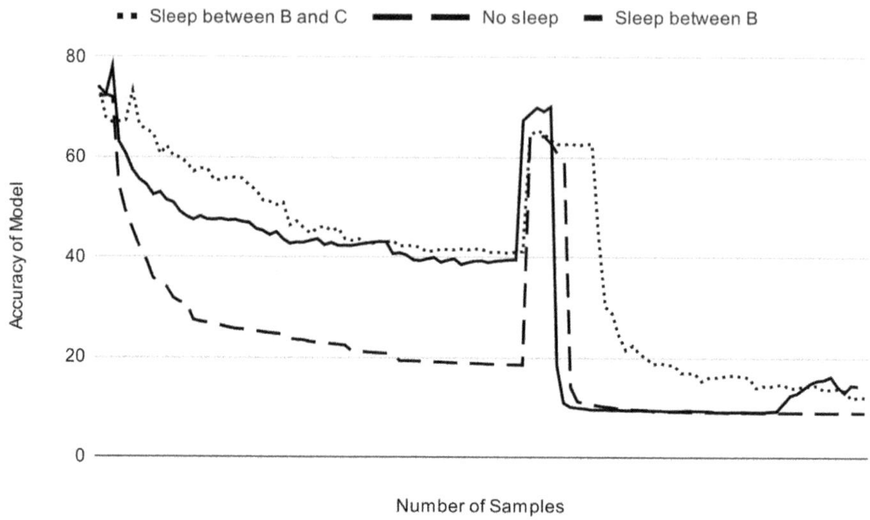

Figure 4.10 Sequential Learning of A Followed by B Followed by C.

Table 4.2 Comparison of Performance of Models with and without Sleep

Parameter	*Without Sleep*	*With Sleep*
Accuracy at 50% samples of B	Accuracy drops catastrophically at a point	Accuracy drops but is greater than accuracy without sleep
Accuracy at 100% training	Accuracy at 25%	Accuracy at 25% to 50% (Upto 2x improvement)
Recall of class B	High	High
Recall of class A	Drops along with accuracy	Drops along with accuracy but at times is greater than without sleep
Catastrophic forgetting	Higher	Lower

4.6 CONCLUSION

In this study, MLPs are used for training data. Various cases and parameters of dream sleep are discussed. The rate of catastrophic forgetting is reduced to a certain extent by the use of reverse learning dream sleep theory. However, this is subject to various hyper-parameters and data characteristics. Based on the results from this study, this idea can be expanded to more complex forms of networks like CNN, RNN (Recurrent Neural Network) and even those networks which closely mimic the structure of the brain, like Hopfield nets or spiking neural networks.

The proposed work presents a novel dream sleep algorithm for multi-layer perceptrons. This algorithm is the application of the Crick and Mitchiston theory of reorganization in ML models. By introducing the proposed "dream cycles" in between training cycles, the rate of catastrophic interference (forgetting) was found to be reduced. This is because of the fact that such dreaming prevents the brain from becoming overloaded, in other words prevents the model from overfitting on the training dataset. This is done by reducing the unwanted connections between neurons, thus acting as a regularization for the model. This study discusses various aspects of this algorithm, its workings, the intuition behind this algorithm, various parameters associated with it and its relation with currently used regularization methods for multi-layer perceptrons. Based on the experimental findings, it is evident that the proposed work shows promising potential.

Notes

1 One of the four phases of sleep in mammals and birds is rapid eye movement (REM sleep), which is defined by the fast and random movement of the eyes [43]. REM sleep is postulated as one factor responsible for reducing the brain's size [7].
2 Not to be confused with reversal learning, which is...
3 Ponto-geniculo-occipital waves (PGO waves) are distinctive waveforms that resemble the random signals occurring in REM sleep, which according to Crick and Mitchison theory act as candidates for activating random neurons in the brain [21].
4 According to the theory, obsession is a state where the model always predicts only a particular set of values [9].
5 It is observed that introducing sleep between A does not have any significant effect.
6 This idea can be extended to three different datasets, for example, A, B, and C—while intermixing sleep between set B and C training.
7 Increasing sleep beyond 20 has little effect on improving the performance of the model, but increases the processing required.
8 In applications where data can be stored, interleaved learning or other methods already in practise can be used in a much more effective manner [12, 13, 32].
9 Similar results have been obtained for multi-epoch training as well.

REFERENCES

[1] Aatmaj-zephyr/ann4j: Artificial neural networks for java this package provides object oriented neural networks for making explainable networks. Object oriented net- work structure is helpful for observing each and every element the model. This package is developed for xai research and development. https://github.com/Aatmaj-Zephyr/ANN4j/tree/main. (Accessed on 12/11/2022).

[2] Ai uses artificial sleep to learn new task without forgetting the last — new scientist. www.newscientist.com/article/2346597-ai-uses-artificial-sleep-to-learn-new-task-without-forgetting-the-last/. (Accessed on 12/11/2022).

[3] How artificial neural networks paved the way for a dramatic new theory of dreams—discover magazine. www. discovermagazine.com/mind/how-artificial-neural-networks-paved-the-way-for-a-dramatic-new-theory-of (Accessed on 12/11/2022).

[4] Researchers made an ai whose performance increases if they let it sleep and dream: Sciencealert. www.sciencealert.com/neural-networks-performance-increases-if-they-re-allowed-to-sleep-and-dream (Accessed on 12/11/2022).

[5] Dmitry I Antonov. Noise data impact on catastrophic interference in spiking neural networks. In AIP Conference Proceedings, volume 2466, page 070003. AIP Publishing LLC, 2022.

[6] George A Christos. Investigation of the crick-mitchison reverse-learning dream sleep hypothesis in a dynamical setting. *Neural Networks*, 9(3):427–434, 1996.

[7] Francis Crick. Neural networks and rem sleep. *Bioscience Reports*, 8(6):531–535, 1988.

[8] Francis Crick and Graeme Mitchison. The function of dream sleep. *Nature*, 304(5922):111–114, 1983.

[9] Francis Crick and Graeme Mitchison. Rem sleep and neural nets. *Behavioural Brain Research*, 69(1-2):147–155, 1995.

[10] Alberto Fachechi, Elena Agliari, and Adriano Barra. Dreaming neural networks: forget- ting spurious memories and reinforcing pure ones. *Neural Networks*, 112:24–40, 2019.

[11] Robert M French. Catastrophic forgetting in connectionist networks. *Trends in Cognitive Sciences*, 3(4):128–135, 1999.

[12] Robert M French. Dynamically constraining connectionist networks to produce dis- tributed, orthogonal representations to reduce catastrophic interference. In Proceedings of the Sixteenth Annual Conference of the Cognitive Science Society, pages 335–340. Routledge, 2019.

[13] galoosh33 (https://stats.stackexchange.com/users/119749/galoosh33). How to avoid 39;catastrophic forgetting39;? Cross Validated. URL:https://stats.stackexchange.com/q/314712 (version: 2017-11-24).

[14] Dong-yuan Ge, Xi-fan Yao, Wen-jiang Xiang, Xue-jun Wen, and En-chen Liu. Design of high accuracy detector for mnist handwritten digit recognition based on convolutional neural network. In 2019 12th International Conference on Intelligent Computation Technology and Automation (ICICTA), pages 658–662, 2019.

[15] Ryan Golden, Jean Erik Delanois, Pavel Sanda, and Maxim Bazhenov. Sleep prevents catastrophic forgetting in spiking neural networks by forming joint synaptic weight representations. bioRxiv, page 688622, 2020.

[16] Ryan Golden, Jean Erik Delanois, Pavel Sanda, and Maxim Bazhenov. Sleep prevents catastrophic forgetting in spiking neural networks by forming a joint synaptic weight representation. *PLOS Computational Biology*, 18(11):e1010628, 2022.

[17] Oscar C Gonzalez, Yury Sokolov, Giri P Krishnan, Jean Erik Delanois, and Maxim Bazhenov. Can sleep protect memories from catastrophic forgetting? *Elife*, 9, 2020.

[18] Ian J Goodfellow, Mehdi Mirza, Da Xiao, Aaron Courville, and Yoshua Bengio. An empirical investigation of catastrophic forgetting in gradient-based neural networks. arXiv preprint arXiv:1312.6211, pages 3521–3526, 2013.

[19] Jarrod A Gott, David TJ Liley, and J Allan Hobson. Towards a functional understanding of pgo waves. *Frontiers in Human Neuroscience*, 11(89), 1-12, 2017.

[20] Jay Gupta. *Going beyond 99% mnist handwritten digits recognition — by jay gupta — towards data science*. https://towardsdatascience.com/going-beyond-99-mnist-handwritten-digits-recognition-cfff96337392. (Accessed on 12/12/2022).

[21] A Hassard. Reverse learning and the physiological basis of eye movement desensitization. *Medical Hypotheses*, 47(4):277–282, 1996.

[22] Tyler L Hayes, Kushal Kafle, Robik Shrestha, Manoj Acharya, and Christopher Kanan. Remind your neural network to prevent catastrophic forgetting. In European Conference on Computer Vision, pages 466–483. Springer, 2020.

[23] hmkcode. *Backpropagation step by step*. https://hmkcode.com/ai/back propagation-step-by-step/. (Accessed on 12/11/2022).

[24] Erik Hoel. The overfitted brain: Dreams evolved to assist generalization. *Patterns*, 2(5):100244, 2021.

[25] Shubham Jain. *Regularization in deep learning with python code*. www.analyticsvidhya.com/blog/2018/04/fundamentals-deep-learning-regular ization-techniques/, Feb 2024. (Accessed on 03/24/2024).

[26] Ronald Kemker, Marc McClure, Angelina Abitino, Tyler Hayes, and Christopher Kanan. Measuring catastrophic forgetting in neural networks. In Proceedings of the AAAI Conference on Artificial Intelligence, volume 32, 2018.

[27] James Kirkpatrick, Razvan Pascanu, Neil Rabinowitz, Joel Veness, Guillaume Desjardins, Andrei A Rusu, Kieran Milan, John Quan, Tiago Ramalho, Agnieszka Grabska-Barwinska, et al. Overcoming catastrophic forgetting in neural networks. *Proceedings of the National Academy of Sciences*, 114(13):3521–3526, 2017.

[28] Johannes Lehtonen. *The theory of crick and mitchison concerning the function of dream sleep: A psychoanalytic comment*. 1988.

[29] Wei Li, Lei Ma, Guang Yang, and Wen-Biao Gan. Rem sleep selectively prunes and maintains new synapses in development and learning. *Nature Neuroscience*, 20(3):427–437, 2017.

[30] Andrew S Lim, Andres M Lozano, Elena Moro, Clement Hamani, William D Hutchison, Jonathan O Dostrovsky, Anthony E Lang, Richard A Wennberg, and Brian J Murray. Characterization of rem-sleep associated ponto-geniculo-occipital waves in the human pons. *Sleep*, 30(7):823–827, 2007.

[31] Jaya Krishna Mandivarapu, Blake Camp, and Rolando Estrada. Self-net: Lifelong learning via continual self-modeling. *Frontiers in Artificial Intelligence*, 3(19), 2020.

[32] Michael McCloskey and Neal J Cohen. Catastrophic interference in connectionist net- works: The sequential learning problem. In *Psychology of learning and motivation*, volume 24, pages 109–165. Elsevier, 1989.

[33] Aliaksei Mikhailiuk. *Deep lifelong learning drawing inspiration from the human brain—by aliaksei mikhailiuk—towards data science.* https://towardsdatascience.com/deep-lifelong-learning-drawing-inspiration-from-the-human-brain-c4518a2f4fb9. (Accessed on 12/12/2022).

[34] Reza Moradi, Reza Berangi, and Behrouz Minaei. A survey of regularization strategies for deep models. *Artificial Intelligence Review*, 53(6):3947–3986, 2020.

[35] German I Parisi, Ronald Kemker, Jose L Part, Christopher Kanan, and Stefan Wermter. Continual lifelong learning with neural networks: A review. *Neural Networks*, 113:54–71, 2019.

[36] Anthony Robins. Catastrophic forgetting, rehearsal and pseudorehearsal. *Connection Science*, 7(2):123–146, 1995.

[37] Grant Sanderson. *3blue1brown—but what is a neural network*? www.3blue1brown.com/lessons/neural-networks. (Accessed on 12/11/2022).

[38] Timothy Tadros, Giri Krishnan, Ramyaa Ramyaa, and Maxim Bazhenov. Biologically inspired sleep algorithm for reducing catastrophic forgetting in neural networks. In Proceedings of the AAAI Conference on Artificial Intelligence, volume 34, pages 13933–13934, 2020.

[39] Yijing Watkins, Edward Kim, Andrew Sornborger, and Garrett T Kenyon. Using sinusoidally-modulated noise as a surrogate for slow-wave sleep to accomplish stable unsupervised dictionary learning in a spike-based sparse coding model. In Proceedings of the IEEE/CVF Conference on Computer Vision and Pattern Recognition Workshops, pages 360–361, 2020.

[40] Wikipedia. *Catastrophic interference—wikipedia.* https://en.wikipedia.org/wiki/Catastrophic˙interferencecite˙note-Ratcliff1990-2. (Accessed on 12/11/2022).

[41] Wikipedia. *Mnist database—wikipedia.* https://en.wikipedia.org/wiki/MNIST˙databasecite˙note-1. (Accessed on 12/11/2022).

[42] Wikipedia. *Pgo waves—wikipedia.* https://en.wikipedia.org/wiki/PGO˙wavescite˙note-:0-1. (Accessed on 12/11/2022).

[43] Wikipedia. *Rapid eye movement sleep wikipedia.* https://en.wikipedia.org/wiki/Rapid˙eye˙movement˙sleep. (Accessed on 12/11/2022).

[44] Han Xiao, Kashif Rasul, and Roland Vollgraf. Fashion-mnist: A novel image dataset for benchmarking machine learning algorithms. arXiv preprint arXiv:1708.07747, 2017.

Chapter 5

Amalgamation of Threat Intelligence for Cognitive Ransomware Detection in 5G Networks

Om Bidve, Radhika Gaikwad, Pragati Nehete, and Rachana Patil

5.1 INTRODUCTION

The ever-evolving landscape of cyber threats necessitates the development of increasingly sophisticated security solutions. Ransomware, a malicious software program that encrypts a victim's data and demands a ransom for decryption, has become a major weapon in the cybercriminal's arsenal. With the advent of fifth generation cellular/mobile network technology (which will be referred to as 5G for the remainder of this chapter), characterized by high speeds, huge capacity and interconnectivity, concerns about ransomware attacks are heightened. Traditional security methods, which often rely on signature-based detection, cannot keep pace with the rapidly changing tactics of ransomware developers.

This chapter looks at the potential of cognitive ransomware detection as a novel approach to securing 5G networks. Cognitive security, a branch of artificial intelligence (AI) applied to cybersecurity, offers a promising way to proactively detect and mitigate ransomware threats. By using machine learning (ML) algorithms and natural language processing techniques, cognitive systems can analyze network traffic patterns, user behavior and communication anomalies to detect early signs of a ransomware attack, even if the malware itself is unknown.

In terms of ransomware attacks, 5G networks are more decentralized in design than earlier generations, which can make them harder to secure. They make extensive use of edge computing, which processes data closer to the source instead of centrally located in a data center. Attacks by ransomware that target edge devices or nodes might interfere with network and local processes. There are vulnerable Internet of Things (IoT) devices that have the potential to be ports of entry for ransomware, which might propagate throughout the network and impair service delivery. Network slicing—the operation of several virtual networks on a single physical infrastructure—is made possible by 5G. Ransomware has the ability to target certain businesses or services, such as healthcare, autonomous vehicles, or smart grids, and disrupt those sectors or services. Due to the significant financial

DOI: 10.1201/9781003501220-5

and reputational risks involved in network outages or service interruptions, telecommunications companies providing 5G services would be more inclined to pay ransoms, making them desirable targets for attacks involving ransomware.

This article examines the specific vulnerabilities of 5G networks to ransomware attacks. It examines the limitations of traditional security methods and how cognitive ransomware detection can address these shortcomings. It also discusses the key features and potential benefits of cognitive systems in the context of 5G network security. By critically analyzing existing research and exploring potential applications, this chapter aims to contribute to ongoing efforts to protect 5G networks from the pervasive threat of ransomware.

5.2 IMPORTANCE OF COGNITIVE COMPUTING IN NETWORKS

Cognitive computing, followed by cognitive technology, will fundamentally change the practice and education of engineers and the engineering systems of the future. They are key drivers for the development of automated capabilities and cognitive goods that are smarter than those on the market today.

The term "cognitive computing" describes the development of computer systems that are modeled on the human brain, capable of processing natural language, learning through experience, interacting naturally with humans and making decisions based on their knowledge. Any system that uses cognitive computing is a learning system. Embedded data analysis and automated management as well as data-centered architectures in which the processor, memory, switching, and storage are positioned ever closer to the data are integrated. Its approach to processing enormous amounts of data is neither deterministic nor linear.

Cognitive computing and cognitive technologies can be seen as the third phase of AI evolution, from traditional AI to general artificial intelligence (AGI) and cognitive systems [1]. In the programmable era, computers essentially process a series of if-then-what equations, while cognitive systems learn, adapt, and eventually hypothesize and propose answers. With the advent of data becoming bigger, faster and more diverse by the day, cognitive computing systems are now being used to extract knowledge from data as experience and then generalize what they have learned to new situations. They tap into the insights that the new wealth of generated data harbors. Delivering these capabilities requires a fundamental shift in the way advances in data processing have been achieved for decades.

The merging of human knowledge with advanced data analysis characterizes cognitive computing. Understanding context necessitates extensive data collection and expert involvement. Despite being in its early stages, cognitive computing presents significant market potential and provides effective problem-solving solutions through ML.

Cognitive computing improves 5G networks by maximizing performance, anticipating problems, enhancing communication, automating tasks, and strengthening security. This innovation is essential for developing intelligent, flexible, and resilient future communication infrastructures.

5.3 PURPOSE AND OBJECTIVES

5.3.1 Purpose

This research paper aims to explore the potential of cognitive ransomware detection as a novel approach to strengthen security in 5G networks. The introduction of 5G networks creates new and exciting opportunities for connectivity and real-time interaction, but it also brings with it serious cybersecurity issues. The sheer volume and complexity of 5G infrastructure provides a large attack surface for malicious actors, while the extremely low latency allows for rapid encryption and exploitation once a device has been infected. Ransomware has the potential to cause significant damage to individuals, organizations and vital infrastructure if left unchecked in 5G environments. To stay one step ahead of the threat, intelligent and dynamic security capabilities are required. This study proposes a new framework for proactive threat intelligence integration to support cognitive ransomware detection in 5G network environments. This technique detects zero-day variants of ransomware early and with low false-positive rates by continuously collecting and analyzing threat data from various internal and external sources, detecting indications of compromise and updating trained detection models in real-time. The proposed smart integration strategy tightly integrates ransomware mitigation capabilities into 5G designs, ensuring the availability and reliability of 5G services even in the face of constantly evolving threats.

5.3.2 Objectives

1. Identify the specific vulnerabilities of 5G networks that make them susceptible to ransomware attacks.
2. Analyze the limitations of traditional security methods in the face of evolving ransomware threats.
3. Explore the capabilities and potential benefits of cognitive ransomware detection systems in the context of 5G network security.
4. Evaluate how cognitive detection can address the shortcomings of traditional methods and improve the overall resilience of the network against ransomware attacks.

5. Contribute to the growth of knowledge and development of effective strategies to mitigate ransomware threats in the evolving landscape of 5G networks.

5.4 MOTIVATION

The heightened susceptibility and reduced latency of 5G networks pose considerable cybersecurity challenges, despite their potential. This research is dedicated to detecting ransomware in 5G by leveraging cognitive computing to fill a crucial gap. Our objective is to strengthen 5G security and durability by conducting thorough testing and validation [2].

With our goal in starting this research project, we aim to equip participants in the 5G ecosystem with the information and resources they need to protect themselves from new cyberattacks. Our study has the potential to significantly impact the security and reliability of 5G networks in a future where connectivity continues to grow through cognitive computing, in addition to enriching scientific discussions.

5.5 RELATED WORK

The hybrid deep neural network in the proposed malware detection system combines long short-term memory (LSTM) and convolutional neural networks (CNN) to detect malicious software in the context of 5G. Initially, a dataset including benign apps and malware is prepared. Static and dynamic methods are employed to extract features such as operation codes, application programming interface (API) calls, and permissions. The LSTM component captures temporal dependencies, while the CNN component identifies spatial patterns. Subsequently, a fully connected neural network layer is utilized for result classification. Techniques like regularization and data augmentation are implemented to enhance generalization and prevent overfitting, and optimization is carried out using algorithms like Adam or Stochastic Gradient Descent. The model's performance is assessed using metrics such as accuracy, precision, recall, and F1-score. [3]

The cognitive security framework being proposed utilizes deep learning (or DL, which, for the remainder of this chapter, will be called deep learning) to detect intruders in 5G and IoT networks by gathering and analyzing network activity data, extracting packet headers, payload details, and flow statistics. Anomalies and intruders are identified using deep neural networks (DNNs) such as auto-encoders and recurrent neural networks (RNNs). Auto-encoders are trained to recognize regular patterns and identify deviations, while LSTMs are employed to recognize time-dependent trends. The framework is designed to be self-learning and adaptable, continually updating itself with new data and attack patterns. To assess its performance, it is compared to traditional intrusion detection systems and other ML methods using metrics like accuracy, precision, recall, and F1-score. [4].

It has been suggested that the framework for cognitive security utilizes deep learning to detect intruders within 5G and IoT networks. It gathers and analyses data on network activity, extracting information from packet headers, payload details, and flow statistics. Anomalies are detected and intruders are classified using DNNs, particularly auto-encoders and RNNs. Auto-encoders are capable of learning typical patterns to identify anomalies, whereas LSTMs specialize in identifying trends that depend on time. This system is self-learning and adaptable, continuously incorporating new data and attack patterns. Its effectiveness is evaluated by comparing it to conventional intrusion detection systems and other ML approaches using accuracy, precision, recall, and F1-score as metrics. [5].

The fundamentals of cognitive computing are discussed in the initial section of this chapter, emphasizing its capacity to replicate human cognitive processes and acquire knowledge from data without specific instructions. The diverse nature of cognitive computing, which encompasses AI, learning algorithms, neurology, and cognitive psychology, is emphasized. The potential for intelligent computing to transform decision-making in various industries such as manufacturing, transportation, healthcare, and finance is a key focus. This is achieved by analyzing large datasets in real time. The challenges of privacy, ethics, and the necessity for dependable algorithms are also addressed in the study. It delves into how cognitive computing can improve [6].

The current state of 5G security is examined in the chapter, with a focus on important issues and enhancements such as enhanced encryption, improved authentication, and backing for network functions virtualization (NFV) and software defined networks (SDN). The chapter explores the use of protocols like 5G Unified Data Management (UDM) and the 5G Authentication and Key Agreement (5G-AKA) protocol for bidirectional authentication. Even with advancements, challenges like distributed denial of service (DDoS) attacks endure, necessitating efficient traffic control and filtering. It is crucial to employ secure virtualization and control via SDN and NFV to prevent unauthorized entry. The integration of 5G with edge computing and IoT calls for robust access control and secure communication. Additionally, the chapter stresses the significance of safeguarding user privacy and adhering to data protection regulations. [7].

The security of 5G is analyzed, with a focus on vulnerabilities in its architecture, user equipment, radio access networks, and core components. Security protocols like the 5G UDM system and 5G-AKA protocol are discussed. Techniques such as fuzzing and penetration testing are utilized to identify any weaknesses. The analysis also scrutinizes the protections in the core network and suggests potential improvements. The integration of 5G with edge computing and IoT is examined, emphasizing the importance of robust access controls and secure communication. Additionally, privacy concerns are addressed, with a focus on protecting user privacy and adhering to data protection regulations. [8].

Sixth generation cellular/mobile technology (6G) networks are confronted with heightened cybersecurity vulnerabilities such as ransomware and malware. ML (which, as a reminder to readers, is short for machine learning) and deep learning present promising solutions for the detection and prevention of cyberattacks. These advanced techniques necessitate extensive network traffic data for effective training. Ongoing research concentrates on ML-based intrusion detection systems tailored for 6G networks, which utilize packet headers and behavioral patterns. Additionally, the utilization of feature selection and ensemble learning serves to bolster accuracy. Deep learning methods, such as CNNs and RNNs, excel in identifying intricate threats within raw data. Furthermore, generative adversarial networks (GANs) are leveraged to fabricate artificial malware, thereby enhancing training processes. The deployment of ML or deep learning models at network edges not only facilitates real-time threat detection and minimizes latency but also safeguards data privacy through edge intelligence and federated learning. [9].

Spectrum scarcity and the limitations of traditional management make opportunistic spectrum access (OSA) and cognitive radio necessary. However, they encounter difficulties in identifying, allocating, and managing interference in the spectrum. An OSA framework based on deep learning, utilizing CNNs, RNNs, and deep reinforcement learning (DRL), can effectively tackle these issues in the context of 5G/B5G (Beyond-5G). DRL is employed to optimize spectrum access according to user requirements and the surrounding environment. Despite challenges such as model interpretability, transfer learning, collaborative computing, and explainable AI can be helpful in the implementation process [10]. The integration of this framework with millimeter wave (mmWave) and multiple-input and multiple-output (MIMO) technologies improves spectrum utilization in 5G/B5G networks [11].

The security of 5G networks is crucial in complex legal environments. Studies concentrate on strong, flexible, and compatible security systems, such as access control, verification, privacy, and blockchain for managing trust. Encryption and secure computation methods guarantee safe data transfer and cooperation [12].

5.6 ISSUES AND CHALLENGES IN RANSOMWARE DETECTION IN NETWORKS

5.6.1 High Data Rates and Low Latency

Detecting ransomware faces two main obstacles: (1) dealing with large amounts of data and (2) adhering to strict time constraints. It is essential to analyze network traffic in real-time in order to detect ransomware early and efficiently. The computational processing required to identify different types of ransomwares from vast amounts of data is quite demanding. Furthermore, the rapid encryption carried out by ransomware makes it crucial to detect and respond to it immediately to avoid significant data loss and operational disruptions. Therefore, a ransomware detection system needs to minimize latency to quickly identify threats and deploy countermeasures. Advanced techniques and strong processing capabilities are needed to overcome these barriers. Network monitoring and analysis tools need to be able to scan signatures for ransom as well as assess massive volumes of data in time, so that they can reliably detect or observe irregular patterns on network traffic. They should also be able to make use of heuristic-based approaches or advanced ML algorithms [13].

In addition, high performance computing resources, such as field programmable arrays, or graphics processing units, could be required to speed up analysis and reduce latency for the detection of ransomware. In order to divide the workload and improve the overall effectiveness of the detection of the ransomware system, distributed architecture and parallel processing can also be used. In addition, in order to keep up with the constantly evolving threat of ransomware, it requires regular updates and adjustments. The detection systems need to be updated with the most recent signatures, behavior trends and analysis approaches if they are to remain effective in detecting and mitigating threats of ransomware.

5.6.2 Massive Device Connectivity

The identification of ransomware is intricate because of the widespread use of interconnected devices, such as IoT, which increases the potential targets for attacks. The presence of diverse devices running different operating systems, security protocols, and with varying resource capacities makes it difficult to create ransomware detection solutions that work universally. Furthermore, the growing connectivity of devices makes it harder to monitor network traffic because ransomware can easily appear as legitimate data flow. The use of advanced techniques like ML and behavioral analysis is made more difficult by the challenge of distinguishing between real and fake network data. Figure 5.1 depicts the evolution of ransomware detection techniques and obstacles from 2020 to 2022. In 2020, there was an exploration of ransomware detection techniques. In 2021, the development of heuristic-based detection algorithms took place. Then, in 2022, the implementation of ML models for ransomware detection occurred. Challenges related to ransomware detection involve the growing complexity of ransomware attacks, scalability issues with current detection systems, and the integration challenges with various IoT devices. Future studies will concentrate on investigating blockchain for decentralized threat intelligence and embracing AI-driven cognitive computing for adaptive defense mechanisms.

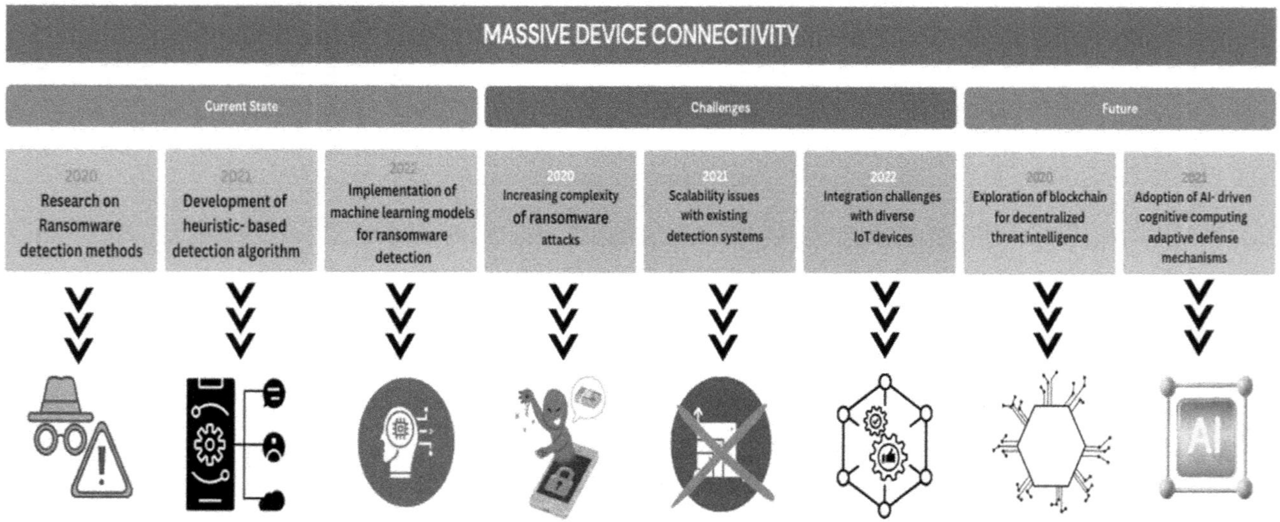

Figure 5.1 Massive Device Connectivity.

5.6.3 Encryption and Anonymity

The detection of ransomware encryption and anonymous communication channels presents major challenges. Traditional signature-based detection is impeded by advanced encryption, requiring the use of behavior analysis and ML for detection. Ransomware operators utilize anonymous communication platforms, making it difficult to trace attack origins and identify threat actors, thus hindering preventative measures. The use of anonymity and encryption further complicates the recovery and response to an incident. In the event of a successful ransomware attack, decrypting affected data or systems can be extremely challenging, even for experienced cybersecurity professionals, if the decryption keys cannot be obtained from the attackers. This may lead to financial losses, operational disruption and major data loss for organizations.

Moreover, as encryption and anonymity algorithms become more complex, the difficulty of detecting and mitigating ransomware will only increase. Security researchers and practitioners need to remain far ahead of the curve in order to successfully combat cyber threats that are constantly evolving within encrypted and anonymous environments, while continuously developing new techniques.

5.6.4 Physical Tampering

Physical tampering is one of the main obstacles to detecting a ransomware attack on networks. Although direct possession of devices or systems can give attackers new channels for infection and possible evasion of detection techniques, the majority of ransomware attacks are started remotely through cyber means. The ability for attackers to circumvent security measures and insert malicious hardware into the systems they are targeting is one of the main challenges with physical interference. An attacker with physical access to the device, for example, could install an infected Universal Serial Bus (USB) drive or other form of storage containing ransomware in order to circumvent network-based detection systems. Physical tampering may also be considered to be the alteration or replacement of hardware, including network connection cards, storage units and even the system as a whole. Attackers may be able to circumvent traditional software detection methods by creating compromised hardware and possibly installing exploitable firmware or backdoor, which will help spread and execute ransomware [14].

Accessing critical infrastructure without authorization physically can result in direct compromise of the system, installation of malware, and circumvention of security protocols. Although barriers and surveillance aid in physical security, insider threats and social engineering still pose risks. It is crucial to adopt a holistic strategy that integrates strong physical security with advanced cybersecurity, encompassing rigorous access controls, employee training, and routine audits to minimize the risks of ransomware.

5.6.5 Resource Constraints

Resource constraints are seriously limiting the effective detection of ransomware on networks. Many people and organizations do not have the security resources, storage space or processing power they need so it can be hard for them to set up effective warning systems against ransomware attacks. One of the main resource constraints is that processing power is scarce, particularly for devices on edge or IoT systems [15]. In order to identify ransomware, such devices are often not equipped with the necessary hardware specifications and processing power for implementing complex security algorithms or analyses of system activities and network traffic in a timely manner. In the same way, storage limitations may make it difficult to keep detailed logs, traffic records, or information backup, all of which are necessary to respond to an incident and to conduct a forensic investigation in the event of a ransomware attack. Organizations with limited storage capacity might be forced to choose between data retention, which could result in the loss of important data that could be useful for detecting and investigating ransomware.

Another major resource limitation is the budget allocated to cybersecurity measures, such as preventing and detecting cyberattacks. Small and medium-sized enterprises often have limited resources, which makes it difficult for them to hire cybersecurity experts, invest in cutting edge security solutions, or put in place extensive training programs for their employees. In addition, security solutions must also be updated and upgraded due to the constantly changing threats of ransomware, which is putting additional pressure on resources. In particular, for companies with limited funding, it may be difficult to keep up with the most recent security updates, threat information and methods of detecting malicious software.

In order to cope with these resource constraints, organizations may need to consider more affordable options such as cloud-based security services, using virtual patching techniques or adopting open-source security tools. Reducing risks and optimizing the effectiveness of the resources at hand in detecting and preventing ransomware can also be helped by setting priorities for security awareness training and developing a robust security culture.

5.6.6 Edge Computing Complexity

Another layer of complexity in the detection of ransomware is the use of edge computing in networks. Decentralization of computing resources and data processing to edge devices, (e.g., IoT sensors, gateways or gateway servers) makes the attack surface for ransomware much more difficult to detect;

it also creates major challenges in detecting them [16]. One of the main challenges is a wide range of devices and systems in relation to edge computation. Other examples of edge devices are resource constrained sensors and embedded systems in addition to more powerful edge servers and gateways. Due to the variability of hardware, software and communications protocols, it is difficult to ensure that all edge computing infrastructure has a unified and thorough method for detection of malicious attacks. Another problem is the scattered nature of computing on edge systems. It is more difficult to monitor and assess potential ransomware attacks due to the various aspects of data processing and decision-making, which affects network traffic and system operation. Traditional centralized security solutions could not work in these decentralized systems, so special security measures had to be put in place at each edge node.

Due to the frequent use of edge devices in harsh or remote locations, they are more vulnerable to physical manipulation or unauthorized access. The malicious actors may be able to physically access these devices and infect them with ransomware or undermine their integrity, in order to avoid network-based detection systems. Furthermore, it may be difficult to obtain timely security updates, (e.g., intelligence about threats and central monitoring and analysis) if connectivity with edge computing environments is limited or absent. This lack of connectivity can make it more difficult to spot and respond in real time to the occurrence of a cyber extortion event. To combat these obstacles, organizations need to implement a multi-layered security strategy that includes centralized monitoring and analysis tools in addition to edge based safety measures. Light safety measures, secure communication protocols and continuous monitoring of edge devices are required to detect efficiently encrypted threats in the computing environment at the edge.

5.6.7 Zero-Day Attacks

Zero-day ransomware attacks are one of the biggest obstacles to detection in networks. A zero-day attack could exploit vulnerabilities and security holes in software or systems that have not been patched or updated before, but are currently without an update. Due to the absence of current ransomware countermeasures, it is very difficult to identify and counter threats that exploit zero-day vulnerabilities. Standard signature-based detection techniques are useless against zero-day malware attacks, which is one of their main problems, because the payload or attack vector is unknown [17]. The ransomware will be able to evade detection and perform its devastating payloads because security solutions based on familiar patterns or signatures are not capable of detecting a new threat.

Additionally, zero-day vulnerabilities can exist across a wide range of software and systems, making them especially dangerous because they can simultaneously affect many people and organizations. These vulnerabilities can be exploited until a patch or security update is released, leaving networks exposed to ransomware attacks in the absence of effect security measures. The stealthy nature of zero-day attacks makes them even more difficult to detect and defend against. The threat actors often use complex tactics to hide their activities and avoid detection, which makes it difficult to detect and respond to attacks in a timely manner. When ransomware is discovered, it may cause significant damage to data encryption, system disruption or data exfiltration. In order to combat zero-day attacks, proactive security measures, such as ongoing vulnerability monitoring, threat intelligence collection and rapid incident response, are needed. Organizations need to prioritize software patching and upgrades to detect anomalies and potential threats that may exploit zero-day vulnerabilities, in addition to implementing advanced security measures such as behavioral analysis, sandboxing and ML detection techniques. In addition, it is vital that information be shared and exchanged within the cybersecurity community with the aim being to promptly detect and mitigate zero-day vulnerabilities and cyber the threats associated with them. By pooling their resources in the analysis and sharing of threat intelligence, organizations can strengthen their overall security posture and be ahead of any possible attack. Figure 5.2 shows the difficulties of identifying ransomware, which involve rapid developments, undiscovered weaknesses, restricted detection approaches, inadequate signatures, versatile techniques, and the requirement for constant monitoring. These obstacles stem from zero-day assaults, making them hard to spot due to their originality.

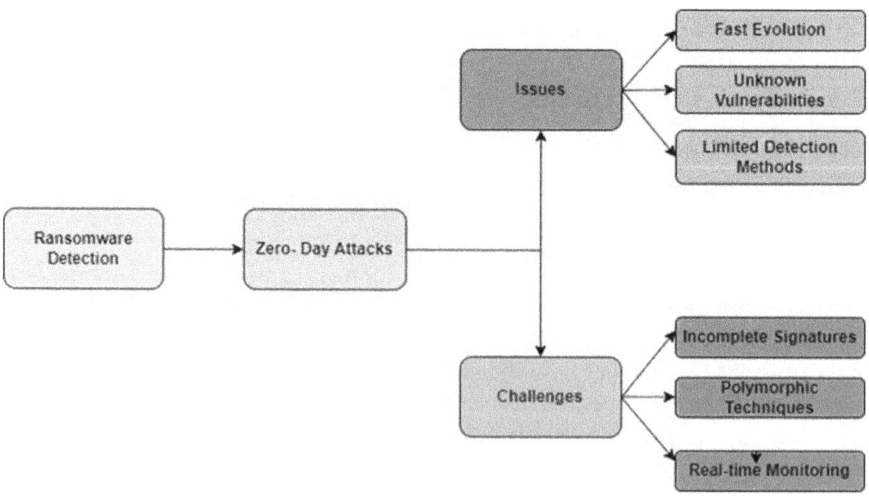

Figure 5.2 Zero-Day Attacks.

5.7 CASE STUDY COGNITIVE RANSOMWARE DETECTION IN 5G NETWORK SECURITY

5.7.1 Smart Home Devices

A cybercriminal can typically access any device on a home network once they gain illegal entry into the network. In the event of this ransom situation, the attackers may make it impossible for a person to use essential features of their house, such as the lights, air conditioning, or appliances. Robot vacuums and smart TV ransomware may seem like minor issues, however, the biggest concern here is that cybercriminals can potentially gain full access to a person's home through any connected device. The land or homeowner would then have to pay in order to gain access to their own property again.

Worse still, is the reality that residents could potentially experience additional violations, even after paying the ransom. The unfortunate scenario which often occurs is that, if the attackers are successful in eliciting a ransom the first time, they might be motivated to come back and demand additional payments. Furthermore, there is no assurance that any data, photos, or audio they gather will not be sold. As a result, the homeowner may become more vulnerable to hacking.

Results from a recent case study investigating ransomware attacks on smart home devices indicate that the digital world (IoT) is a collection of various online-enabled devices and appliances that, because of update issues, frequently lack sufficient security safeguards. With more than 17 billion IoT devices in use worldwide, hacking is a serious risk. Evidence shows that 20–40% of breaches are caused by internal security threats, which means that even smart homes are at risk. Every device is an attack vector since it makes it easier for hackers to gain access using weak passwords or other weaknesses, which could result in the installation of ransomware and network compromise. Parts that are not safeguarded act as invitations for infiltration.

For this reason, it is essential to prioritize home IoT security in order to protect smart gadgets and networks. Another major issue is ignorance—many device owners may be unaware that they've been hacked. Unfortunately, this realization often comes too late, typically in the midst of a ransomware attack. Ransom demands then serve as the first clear sign that their system has been compromised. To explain how ransomware works: typically, a cybercriminal uses it to encrypt a victim's files and then demands payment to unlock them. Attackers prioritize smart homes, interfering with essential operations like lights and appliances, in addition to locking computer folders. For example, they can control thermostats from a distance or set off air conditioning and alarms at the same time. Because smart houses have connected IoT devices, hackers have a lot of control possibilities. These gadgets are open to abuse because they track consumption and allow for remote monitoring and control. As a result, owners of smart homes run certain risks that call for extra caution and security.

5.7.2 Industrial Automation

Proactive ransomware groups including sophisticated nation-state-driven Advanced Persistent Threats (APTs) take advantage of IT/IoT/OT (operational technology) security weaknesses in the current IoT-driven, industrial automation (IA) cyber-threat landscape. Between 2022 and 2023, ransomware attacks in the United States of America (USA) increased by 86%, with 72% of the attacks aimed at manufacturing companies. Critical industrial infrastructures were targeted by Renin-Angiotensin-Aldosterone System (RaaS) at a rate of thirty percent more. There are active ransomware gangs such as Conti and Lockbit. APTs are a serious threat to many different industries as well; in just six years, they have increased their targeting of industrial infrastructure by 300%. XENOTIME, BENTONITE, and CHERNOVITE are well-known IA APT organizations that use strategies such supply chain attacks and zero-day exploits.

Businesses suffer significant financial losses and protracted operational disruptions as a result of these attacks [18]. Because of the possible physical repercussions, cyberattacks on Indian industrial automation call for specialized incident response. Conventional IT methods are insufficient; a combination of IT and OT knowledge is required. The cost of closing this gap is substantial. Integration of cyber insurance is essential to reduce residual risk, notwithstanding obstacles in Indian markets. Small and medium-sized businesses (SMBs) aim for ransomware protection, but they doubt its worth. Cyber insurers have an impact on security procedures, however, depending on risk assessment, they may limit coverage. They cautiously negotiate coverage terms and steer clear of proactive prevention because of privacy concerns. Insurance companies support National Institute of Standards and Technology (NIST) recommendations and emphasize post-event communication for group learning, but they caution against excessive disclosure to reduce the danger of lawsuits.

5.7.3 Healthcare Industry

On Thursday night, March 1, 2024, Rick Pollack, CEO of the American Hospital Association, described the aftermath of a ransomware attack on the nation's largest healthcare payment processor as "the most serious incident of its kind ever directed at a U.S. healthcare organization." According to RansomHub, ALPHV carried out an "exit scam," meaning the gang that staged the cybercrime, stole the ransom money without paying any of the affiliates involved in the attack. RansomHub stated that this refusal to distribute the typical 80% share to the affiliates from the first ALPHV attack led to internal disputes, ultimately causing the group to splinter.

A widely used program that assists healthcare providers with insurance claims and patient payments was rendered inoperative following an attack on Change Healthcare. As a precautionary measure, the company took most of its systems offline to prevent the attack from spreading. UnitedHealth Group disclosed on its website that it had become aware of the attack on February 21 and that the hackers had employed Alphv ransomware. Shortly thereafter, U.S. prescription services were also impacted by additional ransomware incidents. During this period, Change Healthcare experienced a second cyberattack. This new assault, carried out by RansomHub in 2024, followed the ALPHV/BlackCat breach. RansomHub claimed to have obtained over 3,500 source code files from Change Healthcare solutions, personally identifiable information of current U.S. service personnel, medical and insurance records, and 4TB of sensitive data. At the time, they stated that they would release the data if a ransom was not paid [19]. These events underscore Change Healthcare's turbulent start to 2024 and highlight the ongoing challenges that cyberthreats pose to the healthcare industry.

5.8 METHODOLOGY

5.8.1 Evolution of Ransomware in the Context of 3G, 4G and 5G Networks

Ransomware has evolved significantly with the evolution of third generation cellular/mobile network technology (3G), fourth cellular/mobile network technology (4G) and 5G networks. To take advantage of the increased speed, connectivity and capabilities that come with every new generation of cellular networks, ransomware developers have altered their strategies.

The emergence of ransomware, a type of malicious application that hides the victim's information and demands a ransom payment to decrypt it, has posed serious network security problems for decades of wireless communication technologies. In the age of 3G networks, ransomware attacks mainly targeted desktops and servers using loopholes in operating systems as well as programs. In response to the growing popularity of mobile devices, cyber criminals have adapted their strategies and developed a variety of ransomware variations that are tailored for specific mobile platforms which can corrupt websites or install malicious programs on smartphones and tablets. With the introduction of 4G connections and their faster data transfer rates, the threat landscape has changed even more. Campaigns using ransomware have grown increasingly complex, using sophisticated encryption algorithms together with drive-by downloads and phishing to infect devices. With the increasing speed and connectivity of 4G networks, ransomware has been able to spread more rapidly than before, putting consumers and businesses at greater risk. As we transition to 5G networks, it is expected that the possible impact of ransomware will be increased. Thanks to 5G's promised superfast speeds, minimal latency, and massive interconnectedness, cyber criminals will have a new way of attacking. Critical infrastructure systems, smart cities, and IoTs are potential targets for ransomware. This could lead to significant disturbances and, possibly, endanger the safety of citizens. In addition, the emergence of new technologies such as network slicing and edge computing for 5G networks offers a further set of complexities and susceptibilities that can be exploited by ransomware operators. In order to reduce the threat posed by the development of ransomware, a comprehensive strategy is needed. Strong security features such as frequent software updates, sophisticated threat detection and mitigation systems, and user education must be available to all generations of networks. Network operators, device manufacturers and security professionals need to work together in order to stay ahead of the new ransomware strategies as well as ensure effective protection. In order to ensure the safety and resilience of our systems against a constantly evolving ransomware landscape, it is vital that we act proactively and continuously in order to take advantage of the benefits of next generation wireless technologies.

5.8.2 Analysis of Algorithms

A ransomware dataset, containing details on various files and their attributes, is taken from the .csv file. A few details that represent each file in the database are FileName, md5Hash, Machine, DebugRVA, MajorImageVersion, MajorOSVersion, ExportRVA, ExportSize, latVRA, MajorLinkerVersion, MinorLinkerVersion, Numberof Sections, SizeOfStackReserve, DllCharacteristics, ResourceSize, BitcoinAddress and Benign.

A number of files with different names, hashes, machine types, debugging information, versions, export details, linker versions, section numbers, stack reserves, characteristics, resource sizes, bitcoin addresses and a column indicating whether the file is benign or not are included in that dataset.

This dataset is useful for examining and comprehending the traits of ransomware, which may help in identifying and averting ransomware attacks. By analyzing the characteristics of these files, researchers and cyber security experts can better understand how ransomware acts. In the face of such threats, this may improve cybersecurity protection.

The ransomware dataset's values fall between the following ranges for each column:

1. FileName: A string of values representing the file name.
2. md5Hash: A string of values that represent the MD5 hash for this file.
3. Machine: Integer values corresponding to the type of machine, 332 for x86 and 34404 for 69.

4. DebugSize: Integer values representing the size of the debugging information.
5. DebugRVA: Integers representing the respective virtual address of debugging data.
6. MajorImageVersion: Integer values representing the major version of the image.
7. MajorOSVersion An integer value representing the major version of the operating system.
8. ExportRVA: Integer values representing the relative virtual address of the export table.
9. ExportSize: The size of the export table, in integer values.
10. latVRA: integer values representing the relative virtual address of the import address table.
11. MajorLinkerVersion: Integer values representing the major version of the linker.
12. MinorLinkerVersion: Integers representing the minor version of the linker.
13. NumberOf Sections: Integers representing the number of sections in this file.
14. SizeofStackReserve: Integer values representing the size of the stack reserve.
15. DllCharacteristics: Integer values that represent DLL properties.
16. ResourceSize: The size of the resources, expressed in integer values.
17. BitcoinAddresses: Integer values representing the number of addresses associated with bitcoin.
18. Benign: Integer values representing whether the file is benign (0) or malicious (1).

5.8.3 Implementation of Algorithms

The algorithm used in this code is an artificial neural network based on the feedforward variety. The backpropagation algorithm, a supervised learning method, is used to train the neural network by modifying the biases and weights of the network to minimize the function of loss.

Used layers, which are:

1. **Input layer:** There are 10 neurons in the input layer, corresponding to the 10 features in the dataset.
2. **Hidden layer:** There are 64 neurons and the ReLU activation mechanism is activated in this secret layer. The ReLU function, which has been shown to work well in many deep learning applications, is a nonlinear activation mechanism.
3. **Hidden layer:** There are 32 neurons in the second hidden layer, and it also uses ReLU activation.

4. **Output layer:** There is only one neuron with a sigmoid activation function on the output layer. For binary classification problems, the sigmoid function squeezes outputs between 0 and 1, which can be interpreted as a probability of an input belonging to one of the positive groups.

Hyperparameters which can be adjusted are the neural network's depth, the total number of hidden layers and the total number of neurons in each layer to enhance model performance for a given task. The selection of activation functions is another important hyperparameter which may have an impact on model performance.

Simple and widely used, the sequential model in Keras consists of layers stacked on top of each other in a linear fashion; more complex model architectures, such as the Functional API or the model subclass, allow for more flexible and adaptable models. The graph depicted in Figure 5.3 shows the accuracy as a function of time. It begins at 0.8 and maintains a consistent value of 0.8 throughout the entire time period.

Figure 5.4 depicts the accuracy of a sequential model. This is depicted in the graph over 10 epochs, showing a steady increase from 0.1 at epoch 1 to 0.9 at epoch 10, suggesting effective learning by the model.

The data presented in the Pareto chart (see Figure 5.5) indicates that 50% of the total is accounted for by the first category ([1, 5.9]), with the remaining 50% attributed to the second category ((5.9, 10.8]). According to the chart, the first category holds more significance than the second category.

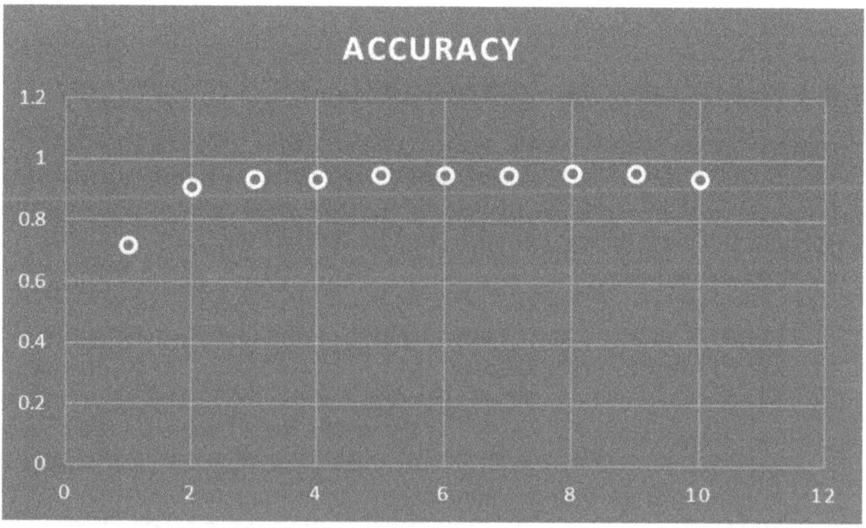

Figure 5.3 Hyperparameter: Accuracy.

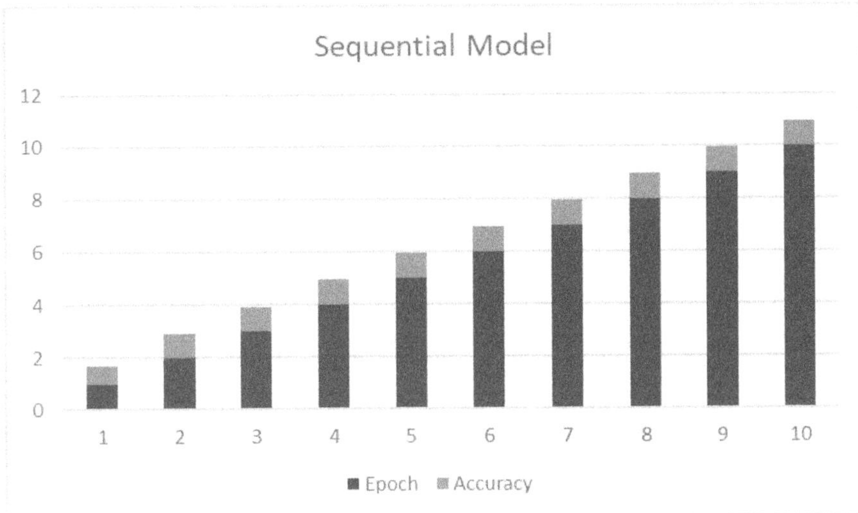

Figure 5.4 Sequential Model.

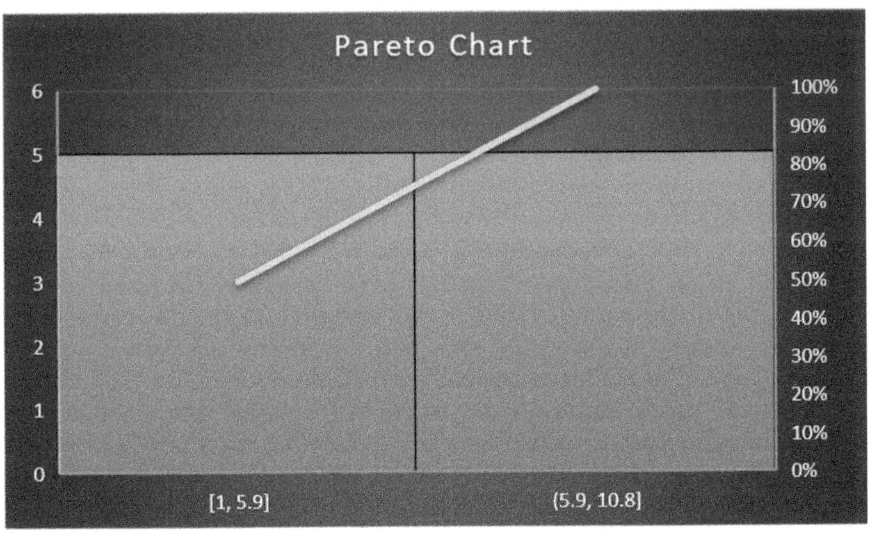

Figure 5.5 Pareto Chart.

With the precision of the validation reaching 97.50% by the end of the training, this output shows that the model is operating well on both the sets used for training and validation. Additionally, the test set evaluation reveals a high test accuracy of 99.00%, indicating good generalization of the model to not-yet-seen data.

A model is often said to have successfully learned the fundamental trends in the data when it shows decreasing loss numbers and increasing accuracy values throughout training. The model appears to have been trained to generate accurate predictions on new, unknown information based on the high validation and test accuracies.

5.8.4 Framework of Proposed System

To detect vulnerabilities, traditionally, the detection systems are planted at specific locations where a mobile network is connected to monitor data at egress ports; correspondingly systems for monitoring the security relies on the physical systems and also interfaces. For mobile networks, network monitoring is an essential tool for network management as it is growing with network and visualization. *threat intelligence for cognitive ransomware detection* would gather the network statistics, traffic and flow samples at different periods and granularities to assess the network's condition for various management activities, including traffic engineering, load balancing, application identification, anomaly detection, network forensics, enforcing service level agreements (SLA), quality of service (QoS), quality of experience (QoE), and network maintenance. Hence the detection system will find the flow of traffic and network to measure the metrics on the basis of intervals and accuracy.

The core transformation in 5G networks is NFV, which allows all network functions to operate as services on commercial off-the-shelf hardware. Additionally, SDN introduces programmable interfaces into networking equipment, decoupling network control from data-forwarding devices and enabling the hardware to be fully programmable. High-end servers have a centralized control plane that can program several network devices at the same time. The SDN control plane can see everything on the network and has more precise control over the packets moving across it. Network monitoring is increased to a higher degree in terms of effectiveness, cost, and complexity as the network is managed by centralized controllers and has programmable interfaces on its component parts. In order to analyze network packet headers, whole packets, or packet payloads, modern networks also use packet sniffing, deep packet inspection (DPI), deep flow inspection (DFI), virus scanners, malware detectors, and other methods. In addition to being used by firewalls, systems for the detection and prevention of intrusions (IDPS), network intrusion detection systems (NIDS), anti-virus scanning appliances, and content filtering appliances, they can be combined with various techniques, such as the areas of ML, behavior analysis, and pattern matching, to identify security breaches (passive security appliances) or stop/block security issues that are discovered (active security appliances).

The architectural options suggested project aims to address the aforementioned problems. For 5G networks, the software-defined monitoring (SDM) architecture has been developed. The SDM architecture for 5G networks is shown in the Figure 5.6.

By using interface extensions similar to OpenFlow as a control interface, switches can provide the metadata, packet information, and flow data needed by security applications. Passive switches monitor and analyze mirrored traffic without interfering with it, whereas active switches handle traffic directly, often passing it through firewalls. The SDM controls the software and hardware security devices and can be integrated with SDN or can remain separate. In case SDN is separate, then it can interact with OpenFlow interface [20]. The control layer is based on SDM, where the application and network infrastructure layers are inserted in between. The operator informs the conductor to deploy a virtualized DPI layer, then the application is set up by utilizing the features to be detected, and lastly, the SDN controller informs the network traffic directly by DPI application function. By using such deployment features, it will be possible to identify the connection's performance characteristics, identify performance issues, and confirm that SLAs are being followed. An SDN architecture is depicted in Figure 5.7, featuring application, control, and network infrastructure layers. Network devices are controlled by the controller using APIs, and security and detection are managed by virtual machines.

5.9 COGNITIVE TECHNIQUES FOR RANSOMWARE DETECTION

There appears to be a possible method of detecting ransomware in networks, thanks to the use of AI and ML algorithms in cognitive techniques. These methods are used for the detection and mitigation of cyber threats based on patterns, behavior or irregularity in system activity and network traffic.

One of the cognitive strategies to detect ransomware is behavior-based analysis. This method trains ML models to identify typical ransomware behavior, such as rapid encryption of files, unusual network connection patterns and suspicious system behavior. These algorithms, by continuously observing and evaluating these activities, can identify anomalies from typical systems and network activity which could indicate the existence of ransomware. Anomaly detection is an additional cognitive tool that uses unsupervised ML algorithms to find odd or suspicious patterns in a system or network activity [21]. These algorithms have been trained to detect typical patterns and any deviations that might hint at the threat of a ransom. When it comes to identifying novel or unidentified ransomware variations

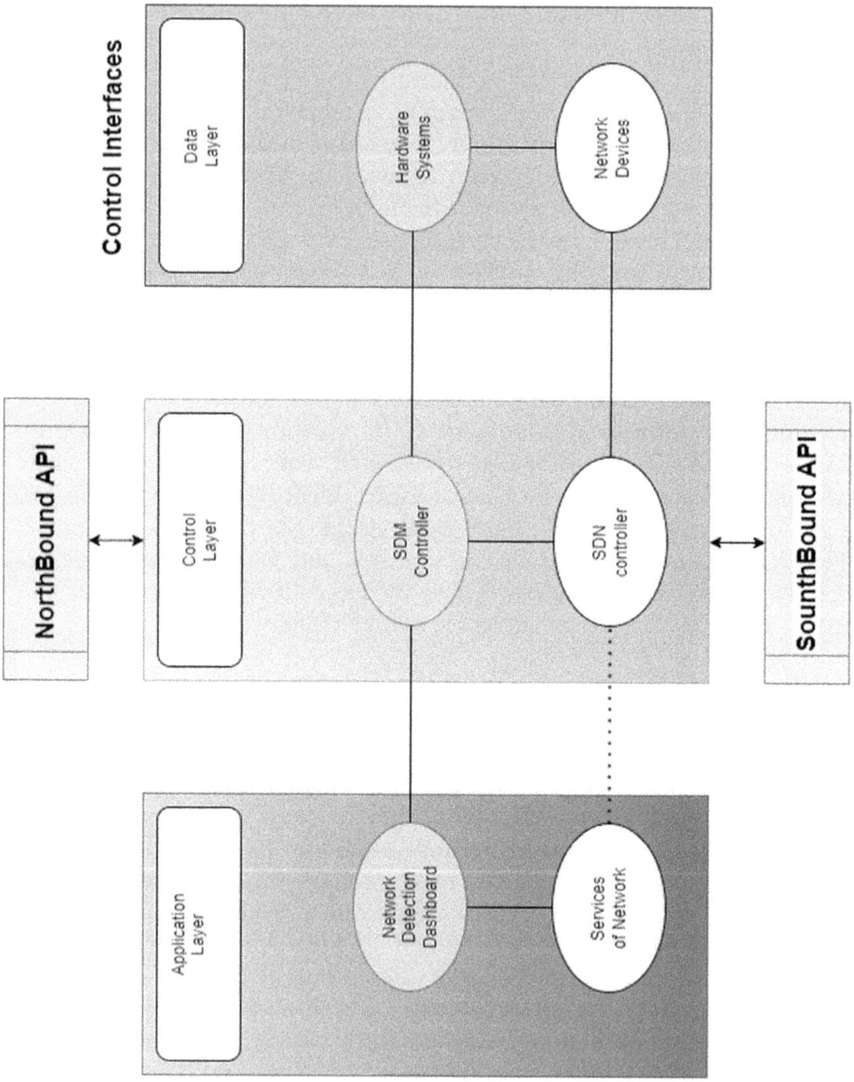

Figure 5.6 Software-Defined Monitoring (SDM) Architecture for 5G Networks.

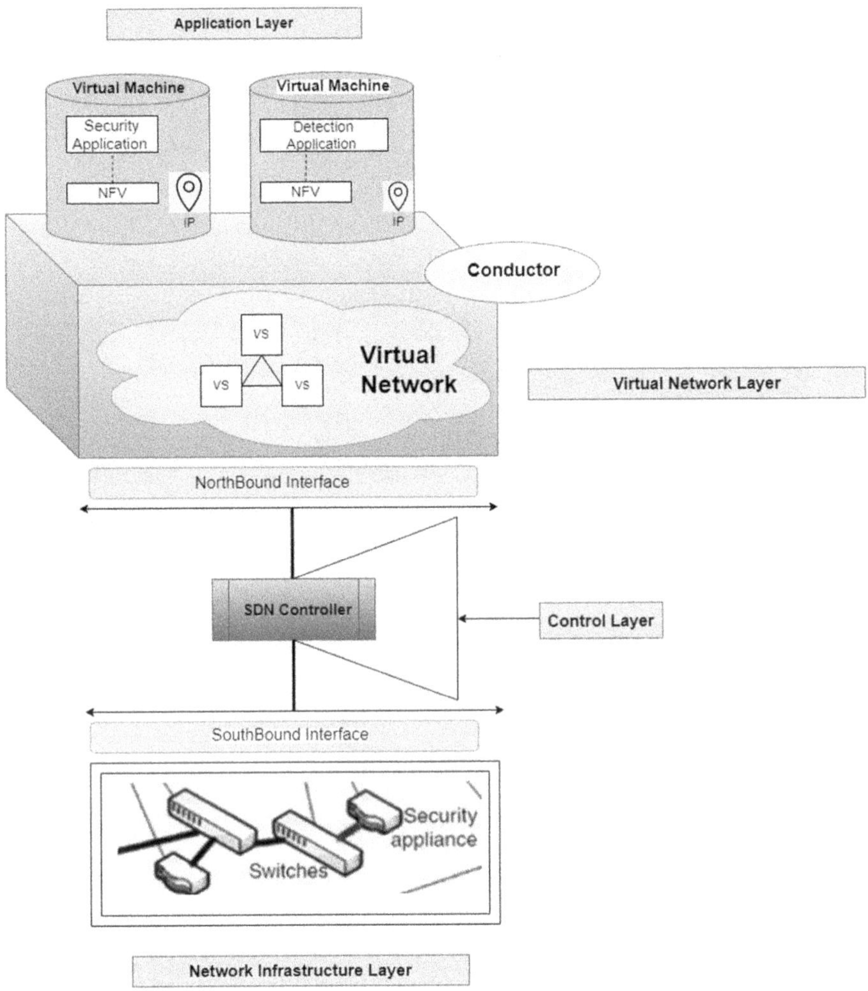

Figure 5.7 Architecture with Application and Network Layer.

that might elude signature-based detection techniques, anomaly detection can be especially helpful. Natural language processing (NLP) tools can also help to detect ransomware by looking for signs of ransomware activity in files or network connections. For example, NLP models can be taught to identify ransom notes or dubious message formats that are frequently used by ransomware operators in order to potentially identify threats before they are able to carry out their destructive payloads. Figure 5.8 depicts the process of using ML to identify ransomware. Initially, data analysis is conducted, which is then followed by ML. ML can be utilized for both

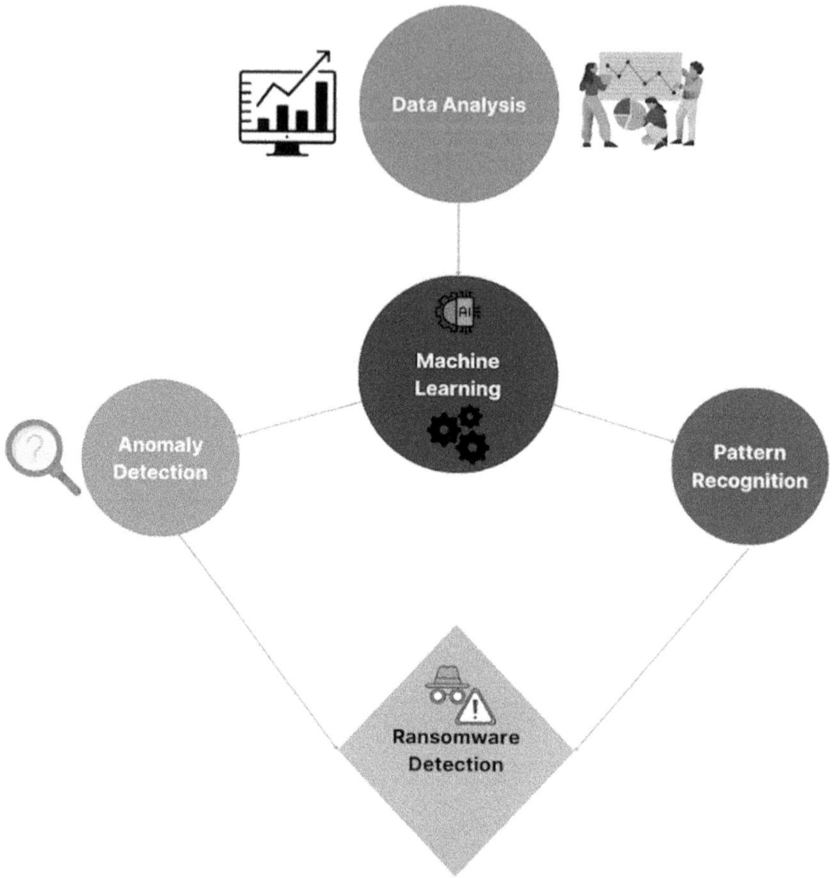

Figure 5.8 Ransomware Detection.

anomaly detection and pattern recognition. These methods can be employed to identify ransomware.

In addition, in combination with other security measures such as dynamic analysis and sandboxing, cognitive approaches can be used to improve the detection of ransomware. These methods make it possible to identify more accurately the threats of ransomware by enabling malicious files or programs to be executed in a controlled environment that allows monitoring and analysis of their behavior. Although cognitive techniques are promising in detecting ransomware, there are drawbacks. In order to train models of ML, there is a need for big volumes of good quality data, but this may be difficult given the new and constantly evolving threat from ransomware. In addition, these methods may require a large amount of computer power and specialist knowledge, which may be difficult for businesses with limited resources [22].

Despite these difficulties, cognitive approaches to ransomware detection offer a potent defense against ransomware assaults in networks by utilizing AI and ML skills to adapt to changing threats.

5.9.1 Tools and Techniques

1. **Python:** Python, which is used for many purposes, including data science and ML, is the computer language used to write the code.
2. **NumPy:** Use the numpy package to perform numerical operations, such as creating a sample dataset and performing matrix operations.
3. **Scikitlearn.sklearn:** The Scikitlearn.sklearn library is used for a variety of purposes, including:
   - *Train_test_split:* This function divides the data into sets for testing and training purposes.
   - *StandardScaler:* The class that is used for standardizing the dataset's features, a common machine learning preparation step.
4. **Keras** is an advanced programming interface for neural network applications based on TensorFlow. This code is used to define, compile, train and evaluate a model of the neural network.
   - *Sequential model:* With a linear layer stack, the Sequential model is a straightforward and widely used model type in Keras.
   - *Dense layers:* The fundamental components of the neural network model, the dense layers are completely connected layers.
   - *The fitting method:* a model is fitted through the fitting method based on training data.
   - *Evaluation method:* The trained model shall be evaluated on the basis of test data and evaluation methods.
5. **Functionalities of activation:** The neural network's hidden layers employ the rectified linear unit (ReLU) activation function. ReLU is a popular activation mechanism in deep learning, due to its efficiency and ease of use.
   - *Sigmoid:* The output layer of the network uses a sigmoid activation function in order to resolve binary classification problems. The sigmoid function, which transposes the output to a value from zero to one, indicates that the input is likely to belong to some positive class.
6. The code shows a simple neural network architecture with two hidden layers; however, in order to maximize the performance of the model, experimentation and tuning are often used to determine the total number of layers, the number of neurons in each layer, and other hyper parameters (such as the learning rate, batch size, and the number of epochs).

5.10 CODE REVIEW

The code implemented is using deep learning algorithms. A sample collection of 1000 samples and 10 characteristics is generated. The dataset is divided into training and test sets using train_test_split from scikit learn. StandardScaler is a tool with standardized features. We are creating a basic feedforward neural network consisting of one input layer, one hidden layer, and one output layer, using the sequential model from Keras. In binary classification, there are one neuron with sigmoid activation on the output layer, 32 neurons that have ReLU activated in the concealed layer, and 64 neurons having ReLU activity within the input layer. The model is constructed using the Adam optimizer and binary cross entropy loss. With a batch size of 32, we train the model on the training data for 10 epochs and validate it on a validation split of the training data. Finally, after assessing the model that has been trained on a test set, we shall print out accurate and lost tests.

Let's break down the code and check its use:

1. **Imports:** The code imports the required modules and libraries, including Keras for neural network model construction and training, numpy for numeric operations, and sklearn to perform data splitting and scaling features.
2. **Create a sample dataset:** Instead of using the real dataset, this code generates a database with 1000 samples and 10 attributes in addition to binary numbers zero and one. Users would substitute that for their own dataset under a real-world scenario.
3. **Divide the dataset:** To ensure repeatability, the dataset is split into sets for training and testing using the train_test_split function in sklearn. A test set with a size of 20% (0.2) and a state that is random of 42 are employed.
4. **Standardize the features:** The training and testing sets are standardized using a standard scaler from sklearn. This is a standard practice in ML to ensure that features have the same scale, so they can improve model performance.
5. **Define a neural network model using the Keras Sequential API:** A neural network model is defined. There are three layers in the model: The input layer is composed of 64 neurons and uses the RuLE mechanism. There are 32 neurons in the hidden layer, which are activated by a ReLU function. One neuron with a sigmoid activation function is available in the output layer, which can be used to solve binary classification problems.
6. **Compile the model:** The binary cross entropy loss function, the Adam optimizer, and the accuracy metric are compiled into the model.
7. **Train model:** The fitting method must be applied in order to train a model using training data. With a batch size of 32 and a validation

split of 20%, the model is trained for 10 epochs (iterations over the full dataset)—the validation set is used to track the model's performance during training.

8. **Assess the model:** Using an evaluation method, which provides loss and accuracy metrics, a trained model shall be assessed on its test set.

5.11 SECURE DEVELOPMENT LIFECYCLE

The basic elements include security monitoring, which involves continuously observing the system for security-related issues.

The core of this process is the following steps in developing security:

1. Keeping an eye on the devices' secure configuration and upgrades is known as "secure device management."
2. Operating system and application patch management is a way of keeping up with the latest security patches for your operating system and applications.
3. Assessment and evaluation of security risks: carrying out regular assessments and evaluations of the system in order to identify possible security risks.

In order to maintain and enhance the overall security of the system, the security development lifecycle is a cycle process in which the different components work together [23]. This strategy will help organizations proactively address security risks across the entire software development and device management lifecycle, instead of responding to security concerns only after a problem has arisen. Figure 5.9 depicts a secure device management system with four primary components: (1) security monitoring, (2) security development lifecycle, (3) operating system and (4) application patch management, and evaluation and inspection of security threats. Each component is linked to the others in a circular manner, illustrating their interconnected and interdependent nature.

5.12 BEST PRACTICES AND GUIDELINES

A thorough strategy to exploit the special advantages of 5G technology is necessary in order to detect cognitive ransomware on mobile networks. To take account of these rules and best practices, here are the following:

1. Implement 5G network slicing to create secure, isolated environments for critical services and constantly monitor slice traffic for any irregularities.
2. Utilize edge computing and AI-driven models to quickly detect and respond to ransomware attacks.

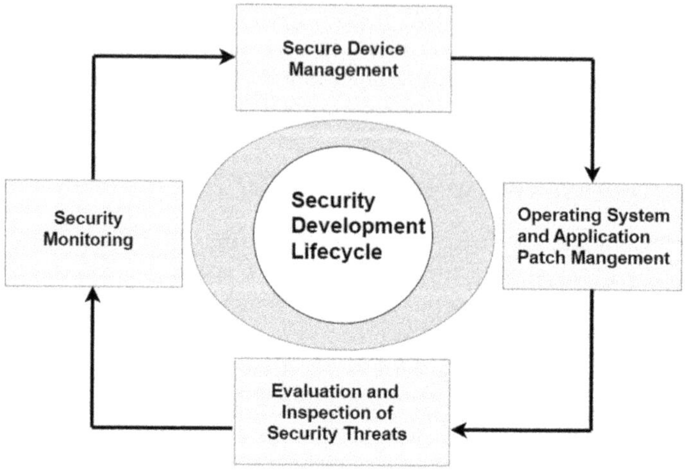

Figure 5.9 Security Development Lifecycle.

3. Enforce strong security measures for 5G devices and applications, such as software updates, secure boot, and application sandboxing.
4. Collect and analyze threat data to forecast ransomware patterns and promptly disseminate threat intelligence updates throughout the 5G network.
5. Safeguard the 5G core and NFV infrastructure with robust encryption, access controls, and monitoring.
6. Encourage the sharing of information and adoption of standardized response protocols among relevant parties to combat ransomware collectively.
7. Educate users about ransomware risks, prevention strategies, and response protocols.
8. Follow industry standards and best practices for 5G security, such as those provided by the 3rd Generation Partnership Project (3GPP) and the International Telecommunication Union (ITU).

To identify the new particular vulnerabilities in 5G network by organizations may identify particular weaknesses in their 5G networks and put in place the necessary safeguards to guard against possible ransomware attacks and other threats by methodically evaluating and fixing these areas. Here are some methods to identify these vulnerabilities:

1. **Make a comprehensive risk assessment:**
   - *Network architecture mapping:* Develop a comprehensive diagram of the 5G network infrastructure, highlighting key

components such as the core network, edge nodes, IoT devices, and virtualized network functions.
   - *Asset inventory:* Conduct a thorough cataloging of all hardware, software, and virtual assets to identify potential targets and entry points for cyber threats.
   - *Threat modeling:* Perform a detailed analysis of potential threats to each component and segment of the network, assessing how vulnerabilities might be exploited by attackers.
2. **Vulnerability scanning and penetration testing:**
   - *Automated vulnerability scanning:* Deploy advanced tools capable of automatically identifying known vulnerabilities across software, configurations, and interfaces within the network. Tools such as Nessus, Qualys, and OpenVAS are widely used for these automated scans.
   - *Penetration testing:* Regularly conduct penetration testing to emulate real-world attack scenarios on the network. This practice helps uncover vulnerabilities that automated tools may overlook, including logic flaws and chained exploits.
3. **Security audits and code reviews:**
   - *Source code analysis:* Execute static and dynamic analyses on network software and applications to uncover coding errors and vulnerabilities, with a particular focus on custom-developed solutions.
   - *Configuration audits:* Assess the configurations of network devices, virtual machines, and software to ensure compliance with security best practices and verify the absence of vulnerabilities such as default credentials or exposed management interfaces.
4. **Network traffic monitoring and analysis:**
   - *Intrusion Detection Systems (IDS)/Intrusion Prevention Systems (IPS):* Implement IDS/IPS solutions to continuously monitor network traffic for suspicious activities and known attack signatures.
   - *Behavioral analytics:* Utilize machine learning-based tools to establish a normal behavior baseline for the network and identify anomalies that may signal security threats.
5. **Protocol analysis and security:**
   - *Protocol analysis:* Review the implementation of network protocols to ensure secure practices, including robust encryption methods, effective authentication mechanisms, and secure error-handling procedures.
   - *Security patches and updates:* Consistently update protocols and software to address vulnerabilities and mitigate newly discovered threats.

6. **Supply chain security management:**
   - *Vendor assessment:* Evaluate the security practices of vendors and partners within the 5G supply chain to ensure adherence to industry standards and best practices.
   - *Software and hardware validation*: Conduct integrity checks on all software and hardware components prior to deployment to confirm they are free from tampering.
   - *Incident response drills:* Execute periodic, scenario-based simulations of security incidents to rigorously test the incident response plan's effectiveness. Assess team performance in identifying, containing, and mitigating threats, while refining communication and coordination protocols. Analyze drill outcomes to improve response strategies and ensure operational readiness for real-world attacks.
7. **Continuous security monitoring and threat intelligence:**
   - *Security Information and Event Management (SIEM):* Deploy a SIEM system to collect, analyze, and correlate security events across the network in real-time.
   - *Threat intelligence:* Integrate real-time threat intelligence feeds to continuously monitor and analyze emerging vulnerabilities, attack vectors, and threat actors targeting 5G networks. Utilize advanced analytics to correlate data from diverse sources, providing actionable insights and timely alerts to proactively defend against evolving global security threats.

5.13 COGNITIVE RANSOMWARE DETECTION IN 5G NETWORK SECURITY

One of the many security risks associated with the 5G technology breakthrough is the constant threat of ransomware attacks. A dangerous cybercrime called ransomware encrypts the victim's files and demands a ransom to unlock them. Traditional signature-based ransomware detection methods are becoming increasingly difficult to use [24]. Cognitive ransomware detection has become a viable method to improve 5G network security. A cognitive approach that scans vast amounts of data for patterns and anomalies that may indicate the presence of ransomware enables the development of advanced ML algorithms and AI. As cognitive systems continuously learn and adapt to new threats, they may be able to proactively respond to such malicious cyberattacks. One of the benefits of this is the ability to detect cognitive ransomware in 5G networks that monitor large volumes of data generated by multiple connected networks devices and services. The volume of data may be too large for traditional security solutions, but cognitive systems can process and analyze this data in real time, allowing potential threats to be quickly identified and mitigated. Radio Access Network (RAN) and User Equipment (UE) are two layers of 5G network architecture that use cognitive approaches. Integrating multiple layers of cognitive security measures that reduce the attack surface and reduce the probability of a successful ransomware attack can create a robust defense [25].

The accuracy and reliability of ML models must be guaranteed. False negative results can cause service interruptions and network vulnerabilities. Because critical information can be exploited or disseminated by an attacker, it is important to carefully assess the security and privacy of data trained with these models. Cognitive ransomware detection is an effective weapon against cyber threats. Using advanced artificial intelligence and machine learning techniques, these systems can proactively and flexibly protect against ransomware. However, the successful implementation of cognitive security measures requires careful consideration of accuracy, privacy, integration and scale. Cognitive ransomware detection is an effective tool to combat 5G network information security cyber threats. Using advanced artificial intelligence and computer programming techniques, these systems are able to provide effective and seamless protection against Ransomware attacks. However, the successful implementation of cognitive security measures must consider issues of accuracy, privacy, integration and scalability.

5.14 COGNITIVE COMPUTING IN CYBERSECURITY

To begin with, cognitive computing, a subset of AI, has deep historical roots dating back several centuries. Since 1974, significant advancements in this field have greatly influenced the development of modern technology. The collaboration of expert scientists and mathematicians led to the development of models that simulate the functioning of the human brain, aiming to replicate its ability to perform complex tasks and solve problems efficiently. Since then, a large number of scientists have contributed to the advancement of what is now known as contemporary AI (which, to remind the reader, stands for artificial intelligence). The core part of cognitive computing is processing data,

A key factor in cognitive computing in cyber security is the ability to process and analyze vast amounts of data from many different sources, such as network traffic, system logs, user behavior patterns or threat intelligence. By applying ML algorithms to this data, cognitive systems can identify trends and anomalies that may indicate malicious activity, potential vulnerabilities or risks. The ability to continuously learn and adapt is a major advantage of cognitive computing. To detect and counter threats, intelligent systems can adapt their models to include new risks and attack vectors. Reliance on rule-based systems and manual updates is reduced due to the ability to learn independently.

Cognitive computing can be applied to several areas of cybersecurity, including:

1. **Threat identification and response:** Recognizing dangers such as malware and illegal access and taking appropriate action to counter them.
2. **Vulnerability management:** The proactive identification and ranking of software and system flaws.
3. **Analyzing user behavior:** Finding irregularities that point to account compromise or malicious activities.
4. **Fraud detection:** Using transaction data analysis, financial crimes are found.

Assuring model accuracy, protecting data privacy, and successfully integrating technology across teams are obstacles, too.

5.15 THE FRAMEWORK FOR PROPOSED SYSTEM

The foundation for cognitive ransomware detection in a 5G network consists of several interrelated components, each of which is required to provide robust protection against ransomware threats and attacks. Through state-of-the-art ML and AI techniques, this framework provides intelligent, predictive security solutions that are flexible.

1. **Data acquisition and processing:** This is the first step in collecting and preparing data from multiple sources in a 5G network. Infrastructure data includes user behavior patterns, system logs, threat intelligence data and network traffic data. Pre-processing techniques such as filtering, normalization and feature extraction are used to ensure that the data is properly formatted.
2. **Cognitive analysis engine:** This is the second stage where the preprocessed data is analyzed using AI and advanced machine learning algorithms. A component responsible for identifying trends, anomalies, and warning signs of potential ransomware threats.
3. **Threat intelligence integration:** Collecting and analyzing data from external sources such as industry reports, threat intelligence platforms, and cybersecurity databases is part of this distinctive architecture. The cognitive system uses this intelligence to expand its knowledge.
4. **Risk assessment and prioritization:** The cognitive system evaluates perceived weaknesses or risks. It assesses their level of risk based on variables such as criticality, impact and probability of occurrence. This risk assessment allows security measures and resource allocation to be prioritized.

5. Appropriate response and mitigation techniques are proposed for the next step. This may include automated processes such as defining an incident response protocol, isolating affected systems, or blocking offending traffic.
6. **Continuous monitoring and feedback loop:** This framework uses both response loops and continuous monitoring to ensure efficiency and flexibility. They monitor the results of implemented security measures and use this information to improve cognitive models and decision-making processes.
7. **Integration with current security architecture:** The cognitive ransomware detection system integrates seamlessly with existing security measures in 5G networks. This creates a layered defense against ransomware, minimizing disruption and allowing collaboration between the new system and traditional security tools. This improves 5G network security by offering a more comprehensive and adaptable approach to ransomware detection and response.

5.16 CHALLENGES AND VULNERABILITIES IN 5G NETWORKS

In addition to significant improvements in speed, delay and connectivity, 5G networks also provide new vulnerabilities and challenges that need to be addressed. Increased threat surface, by adding new technologies to 5G networks, such as edge computing, software defined networking, virtualization, and network function virtualization NFV, the threat surface will be increased, which will make it easier for hackers to penetrate the network.

1. **Vulnerability related to virtualization:** Another security threat is created by the use of virtualization techniques, such as network slicers and virtual networking functions (VNFs). Poor isolation of virtual resources or risk factors in the virtualization layer could lead to an unauthorized access or crosslinking attack.
2. **Distributed architecture:** It is more difficult to continuously protect the complete network infrastructure in 5G networks due to their distributed design, which has components such as ground stations, nodes at the edge, and core network operations dispersed over multiple places.
3. **IoT and enormous connectivity:** There is a greater chance that hacked devices may be exploited as attack entry points or as a component of botnets that operate for distributed denial-of-service (DDoS) assaults due to the vast number of IoT devices anticipated to connect to 5G networks.

4. **Software vulnerability:** As 5G networks are more reliant on software solutions such as SDN and NFV, there is a chance that malicious parties could exploit these weaknesses.
5. **Signal security:** The 5G alert protocols, which are responsible for launching and maintaining communication sessions, can be targeted by attacks like signal overload, hijacking the browser or manipulating them. This could result in interruption of services or data breaches.
6. **Threat from quantum computing:** Threats to the security of cryptographic algorithms used in 5G networks, which require development of quantum resistant encryption techniques, can be jeopardized if quantum computers are developed.
7. The huge amount of data created and transmitted over 5G networks, as well as the integration of a number of services and applications, raises concerns about user safety and privacy.
8. **Security of the supply chain:** Because of the 5G ecosystem's links with several suppliers and suppliers, there is a greater chance that the hacked hardware or software will find its way into the network's infrastructure.
9. **Lack of coordination and standardization:** Disparities in security procedures and standards may arise as a result of the rapid development and deployment of 5G networks across many countries and vendors, making it more difficult to ensure complete safety.

5.17 SECURITY PROTOCOLS IN 5G

A number of security protocols and technologies have been developed in the architecture of 5G networks to address security issues and vulnerabilities. These protocols aim to ensure the availability, confidentiality and integrity of data and services in the 5G ecosystem. Several security protocols used in 5G networks are listed below:

1. **5G Authentication and Key Agreement (AKA):** The most common form of authentication in a 5G network is the protocol. It is responsible for ensuring that the network and the user equipment (UE) authenticate each other. It is also responsible for establishing secure communication routes. It is an evolution of the AKA (Authentication and Key Agreement) protocol of previous generations.
2. The fifth generation 5G PDCP protocol, or Packet Data Convergence Protocol, operates in the PDC layer. It offers a wide range of services that include strong security features such as two-factor authentication, integrity protection and encryption. It ensures the integrity and confidentiality of signals and user data sent through the radio interface.

3. **5G Gateway Route Control Protocol (5G GRC):** The 5G GRC protocol is responsible for secure routing and traffic control in the5G core network. It ensures the confidentiality and integrity of communication between different network services according to the control level messages passed from one service to another. It protects against possible rerouting and traffic manipulation and unwanted access.
4. **Fifth generation non-access stratum (5GNonAccess Stratum NAS):** Signal packets sent between the UE and the core are protected by the 5G protocol when sent back and forth between the UE and the core. to the network. It provides mobility management, authentication and secure deployment methods. Confidentiality and integrity of critical control plane processes are protected.
5. **IPsec and transport layer security (TLS):** IPsec provides secure tunneling and data encryption for the control plane and user plane traffic, while TLS provides secure communication between network functions and management systems.
6. **Subscriber privacy and data protection:** 5G networks have a number of safeguards to address user privacy and data protection concerns, including subscriber identity confidentiality, subscriber privacy mechanisms, and lawful interception capabilities. These measures are intended to protect user data and ensure compliance with legal and privacy standards.
7. **Trusted Execution Environments (TEE):** TEEs are secure, isolated execution environments found within 5G network infrastructure. They are designed to protect sensitive data and the security of critical functions against possible intrusions or vulnerabilities in the primary operating system.

5.18 STUDY OF ALGORITHMS FOR RANSOMWARE DETECTION

The following study explores various algorithms that can be used for effective ransomware detection in the context of 5G networks.

1. **Machine learning-based algorithms (ML):** In the field of cybersecurity, ML algorithms have proven to be powerful tools for complex and dynamic. processes. tools threats like ransomware. These algorithms can analyze large amounts of data, allowing them to identify trends and use their learning models to make intelligent decisions.
2. **Supervised learning:** Labelled data, including ransomware samples, can be used to train algorithms as support vectors for random forests. (SVM) and DNN. These algorithms are trained to distinguish between malicious activities and those that are not. This

enables accurate detection of known and possibly unknown variants of ransomware.

3. **Unsupervised learning:** Algorithms such as clustering techniques such as K means, density-based spatial clustering of applications with noise (DBSCAN) and anomaly detection systems such as Isolation Forest, One Class of SVM can be used to detect deviations from normal system or network behavior. This seems to indicate the presence of ransomware.
4. **Semi Supervised learning:** Algorithms such as self-training and CoTraining can benefit from both labelled and unlabeled data when it comes to situations where there is scarcity, high costs or difficulty obtaining the label information.
5. **Algorithms based on deep learning:** Deep learning is a branch of machine learning that makes use of multi-layered artificial neural networks. It's proven to be a huge success in cybersecurity. These algorithms allow for the automatic extraction of relevant features from raw data, which will improve detection efficiency.
6. **Convolutional neural networks (CNNs):** CNNs can be used to examine structured data, such as binary file signatures and network traffic patterns. In addition, by detecting patterns which indicate the presence of crime, they can be used to identify and decrypt ransomware.
7. **Recurrent Neural Networks (RNNs):** There are several forms of recurrent neural networks, such as gated non-ribosomal peptide synthetases (gated NRPs) and long-term memory. They are very good at deciphering sequential data, such as network packet sequences or system logs. Therefore, over time they can be used to track the propagation of ransomware.
8. Autoencoders are unsupervised deep learning models which can be trained to replicate the input data. They are helpful for detecting anomalies since they can detect deviations from the norm, which may indicate that a computer is infected with ransomware
9. **Hybrid algorithms:** Many techniques are integrated into hybrid algorithms in order to maximize the benefits of each method. They can, for example, combine deep learning with conventional feature engineering or machine learning with rule-based systems.
10. **Behavioral analysis algorithms:** These algorithms are focused on detecting the unique patterns and actions shown in ransomware.
11. **System call sequence analysis and monitoring:** These techniques can detect patterns of behavior indicative of a ransomware attack. Resource exhaustion, data exfiltration and file encryption are some of the cases.
12. **Honeypot analysis:** By placing decoys or honeypots in 5G networks, it is possible to attract attention and record ransomware activity,

which provides important information for developing effective detection algorithms.

13. **Sandbox and dynamic analysis:** When you run a secure sandbox environment for suspicious files or applications, you can identify the characteristics of ransomware by observing its behavior. Changes in patterns, network communication and system resource consumption can be exploited.
14. **Graph-based algorithms:** By modeling the connections and exchanges between different components of a 5G network, graph-based algorithms can be used to detect unusual behavior patterns that may be indicative of ransomware activity.
15. **Graphic neural networks (GNNs):** The behavior and topology of 5G network components can be studied using GNNs. Mysterious patterns that can be considered ransomware are detected through detailed learning.
    - *Belief propagation and message passing:* Using these methods, information can be propagated across the network graph, enabling the detection of synchronized or distributed ransomware activity.

Choosing a ransomware detection algorithm: This depends on several variables, including network specifications, available data sources and available computer resources. In real applications, the detection accuracy and flexibility of these methods can be improved by combining several different algorithms and methods.

5.19 BEHAVIORAL ANALYSIS AND DECISION-MAKING PROCESSES IN COGNITIVE DETECTION

5G networks use AI and ML techniques to detect cognitive ransomware, which analyze complex behavior patterns and provide a clear view of potential threats.

Identification of cognitive ransomware:

1 **Data collection:** Compile a variety of network data for analysis, including sensor, system logs, user behavior, and traffic statistics.
2. **Feature extraction:** Determine salient features suggestive of ransomware activity.
3. **Machine learning (ML):** Use both unsupervised (anomaly detection) and supervised (known ransomware patterns) algorithms.
4. **Contextual analysis:** Recognize intricate malware patterns over time and across machines.
5. **Decision and response:** Recognize dangers and take the necessary precautions (such as traffic blocking and isolation).

5.20 POTENTIAL ATTACK VECTORS FOR 5G

This section explores the potential attack vectors in 5G networks, highlighting the various ways adversaries could exploit vulnerabilities within the system.

1. DDoS attacks, also known as distributed denial-of-service attacks, are more likely to occur on 5G networks due to their increased capacity and connectivity. An adversary adds network resources that create excessive traffic that can cause service interruptions or problems. By launching a distributed denial-of-service attack from compromised devices, these attacks can exploit the large number of connected nodes in the 5G ecosystem.
2. **Radio access network vulnerabilities:** The 5G RAN is responsible for connecting UEs to the core network. Examples of RAN components that may be vulnerable to unauthorized access, data collection, or service interruption are base stations. Vulnerabilities in communication protocols or new 5G radio interfaces can also be exploited by interested parties.
3. **Using network functions virtualization (NFV) and software defined networks (SDN):** As 5G networks rely heavily on SDN and NFV technologies, they are exposed to vulnerabilities in both underlying virtualization platforms and virtualized web services. Adversaries can use these components to launch lateral attacks, disrupt services, or gain unauthorized access by exploiting their incompatibilities or vulnerabilities.
4. **Supply chain attacks:** The 5G network architecture consists of a complex supply chain consisting of hardware and software components from multiple vendors and manufacturers. Weaknesses or malicious code at any point in the supply chain can compromise network security.
5. **Internet of Things (IoT) vulnerabilities:** The number of IoT devices connecting to 5G networks allows attackers access to entry points. Many IoT devices are vulnerable to exploitation because they often have insufficient security features or insufficient software updates.
6. **User privacy and security risks:** Vulnerabilities or lack of security measures may allow unauthorized access to user data, which may compromise users' personal security and opens the door to possible exploitation.
7. **Signal overload attacks:** An attacker's signal overload attack can focus on the new signaling protocols and procedures developed by the 5G network architecture. These attacks aim to use excessive signaling traffic to compromise the control layer of the network.
8. **State actors and Advanced Persistent Threats (APTs):** 5G networks are becoming critical infrastructure for multiple services and sectors.

They can become attractive targets for attackers. These hackers may use a variety of strategies to control access or user actions, such as sophisticated malware, zero-day exploits, and social engineering.

5.21 INTEGRATION METHODS FOR THREAT INTELLIGENCE

Immediate risk removal, intelligence decision making, and proactive threat identification are facilitated by incorporating threats in the security architecture of 5G networks. Here's some of the techniques:

1. **Integration of event management with security information:** From a number of sources, security information and event management (SIEM) data are collected, analyzed, interrogated using SIEM in the organization's IT infrastructure. This integration enables real time threat monitoring, alert production and response to incidents.
2. **Security orchestration, automation, and response integration (SOAR):** By combining different security technologies, procedures, and data sources, SOAR systems are made to automate and streamline security operations. Threat information can easily be incorporated into SOAR platforms. This combination is likely to significantly enhance the effectiveness and efficiency of threat detection and response activities on 5G networks.
3. **Threat Intelligence Platform Integration (TIP):** These are specialized programs that collect, analyze and share threat intelligence from multiple sources. It consists of proprietary research, closed source inputs and open-source data. Organizations can improve their security monitoring and vulnerability management by integrating TIPs into the 5G network security infrastructure.
4. **Threat intelligence platform integration (TIP):** These are specialized programs that collect, analyze and share threat intelligence from multiple sources. from the sources. It consists of proprietary research, closed source inputs and open-source data. By integrating TIP into the 5G network security infrastructure, organizations can improve their security monitoring and vulnerability by gathering threat intelligence. stunning scenery.
5. **Integration of network security devices:** Access threat information by configuring intrusion prevention systems, web application firewalls, and firewalls. With this integration, these devices can instantly block known malicious IP addresses, URLs and attack signatures.
6. **Integration of deployment management and vulnerability management:** Threat intelligence can provide critical information about vulnerabilities affecting different parts of the 5G network infrastructure. This includes all necessary hardware and software. Organizations

can reduce the risk of successful exploitation by prioritizing and accelerating remediation of critical vulnerabilities when they integrate threat intelligence with vulnerability and patch management systems.

7. To effectively integrate threat mitigation, the 5G core network, Radio Access Network (RAN) and UE are just a few of the many components to consider strong data management procedures, such as data quality assurance, privacy and regulatory compliance, or access control mechanisms, should support the integration of threat intelligence to ensure that sensitive data is handled securely and correctly. By leveraging these integration technologies offered by 5G network operators, security teams can improve their situational awareness and stay abreast of evolving threats.

5.22 FUTURE WORK

Subsequent research in this area may investigate how to augment the contextual awareness of ransomware threats in 5G networks through the use of sophisticated threat intelligence methods like sentiment analysis and natural language processing. Furthermore, while protecting data privacy and sovereignty, the development of decentralized and federated learning methodologies may make it possible to train models collaboratively across dispersed network settings. Reducing reliance on centralized detection systems, real-time threat mitigation and response at the network edge may be made possible by expanding cognitive ransomware detection capabilities to edge computing nodes within 5G networks. In addition, investigating the application of blockchain technology for audit trails and immutable threat intelligence sharing may improve the integrity and transparency of threat intelligence data transfers between relevant parties.

By exploring these possibilities, scientists may keep coming up with new ideas and improving the state-of-the-art in cognitive ransomware detection for 5G networks, strengthening the security and resilience of these networks against ever-changing cyberthreats.

5.23 CONCLUSION

As a result, although 5G networks present previously unheard-of chances for real-time communication and connectivity, they also pose serious cybersecurity risks, namely in light of ransomware attacks. This paper presents a novel architecture for proactive integration of threat intelligence that is customized for the complex environment of 5G networks. Our approach uses cutting-edge algorithms and deep learning techniques to create an intelligent ransomware detection system. By carefully examining and optimizing crucial factors, our goal is to improve model efficiency and facilitate the instantaneous detection of ransomware assaults in 5G settings. This study advances computational thinking in 5G networks while simultaneously addressing the pressing need for strong security measures. This seamless integration of cognitive ransomware detection into 5G designs provides the dependability and robustness of 5G services in the face of constantly changing threats. The ultimate goal of this research is to provide stakeholders with the knowledge and resources they need to protect 5G infrastructure's integrity and lessen the growing threat that ransomware attacks offer.

REFERENCES

[1] Modha D. S. et al., (2011). Cognitive computing. *Communications of ACM*, 54, August 2011, 62–71/, http://dl.acm.org/citation.cfm?id=1978559&bnc=1.

[2] Taheri, R., Ghahramani, M., Javidan, R., Shojafar, M., Pooranian, Z., & Conti, M. (2020). Similarity-based android malware detection using hamming distance of static binary features. *Future Generation Computer Systems*, 105, 230–247. https://doi.org/10.1016/j.future.2019.11.034

[3] Lu, Ning, et al. (2021). An efficient combined deep neural network based malware detection framework in 5G environment. *Computer Networks*, 189: 107932.

[4] Lilhore, Umesh Kumar, Dalal, S., & Simaiya, S. (2024). A cognitive security framework for detecting intrusions in IoT and 5G utilizing deep learning. *Computers & Security*, 136: 103560.

[5] Mozo, A., Pastor, A., Karamchandani, A., de la Cal, L., Rivera, D., & Moreno, J. I. (2022). Integration of machine learning-based attack detectors into defensive exercises of a 5G cyber range. *Applied Sciences*, 12, 10349. https://doi.org/10.3390/app122010349

[6] Noor, Ahmed K. (2014). Potential of cognitive computing and cognitive systems. *Open Engineering*, 5(1), 10.1515/eng-2015-0008. https://doi.org/10.1515/eng-2015-0008

[7] Jover, Roger Piqueras. (2019). The current state of affairs in 5G security and the main remaining security challenges. *ArXiv* abs/1904.08394: n. Pag. https://arxiv.org/pdf/1904.08394

[8] Piqueras Jover, R. & Marojevic, V. (2019). Security and protocol exploit analysis of the 5G specifications, *IEEE Access*, 7, 24956–24963, , https://doi.org/10.1109/ACCESS.2019.2899254.

[9] Ankita, A. & Rani, S. (2021). Machine Learning and Deep Learning for Malware and Ransomware Attacks in 6G Network, 2021 Fourth International Conference on Computational Intelligence and Communication Technologies (CCICT), Sonepat, India, pp. 39–44, doi: https://doi.org/10.1109/CCICT53244.2021.00019.

[10] Aldhaheri, Alyazia, et al. (2023). Deep learning for cyber threat detection in IoT networks: A review. *Internet of Things and Cyber-Physical Systems*, 4. 10.1016/j.iotcps.2023.09.003

[11] Ahmed, R., Chen, Y., & Hassan, B. (2021). Deep learning-driven opportunistic spectrum access (OSA) framework for cognitive 5G and beyond 5G (B5G) networks. *Ad Hoc Networks,* 123: 102632.

[12] Jorquera Valero, J. M., Sánchez Sánchez, P. M., Lekidis, A. *et al.* (2022). Design of a security and trust framework for 5G multi-domain scenarios. *Journal of Network and Systems Management*, 30(7). https://doi.org/10.1007/s10922-021-09623-7

[13] Malik, J. & Kaushal, R. (2016). CREDROID: Android malware detection by network traffic analysis, in: *Proceedings of the 1st Acm Workshop on Privacy-Aware Mobile Computing*, pp. 28–36.

[14] Yan L. K. & Yin H. (2012). Droidscope: Seamlessly reconstructing the {OS} and dalvik semantic views for dynamic android malware analysis, in: Presented as Part of the 21st {USENIX} Security Symposium, {USENIX} Security 12.

[15] Wang, Z. & Liu Q. (2020). Review of Android Malware Detection Based on Deep Learning. https://ieeexplore.ieee.org/document/9211502

[16] Kang, B., Yerima, S.Y., Sezer, S., & McLaughlin, K. (2020). N-gram opcode analysis for android malware detection. *International Journal on Cyber Situational Awareness*, 1, 231–255. 10.22619/IJCSA.2016.1001011.

[17] Kim, Y. (2017). Convolutional neural networks for sentence classification. *Proceedings of the 2014 Conference on Empirical Methods in Natural Language Processing*. 10.3115/v1/D14-1181.

[18] Sun, X., Ngueilbaye, A., Luo, K., Cai, Y., Wu, D., & Huang, J. Z. (2024). A scalable and flexible basket analysis system for big transaction data in Spark. *Information Processing & Management*, 61(2), 103577. https://doi.org/10.1016/j.ipm.2023.103577

[19] Ahmed, K., Nadeem, M. I., Zheng, Z., Li, D., Ullah, I., Assam, M., Ghadi, Y. Y., & Mohamed, H. G. (2023). Breaking down linguistic complexities: A structured approach to aspect-based sentiment analysis. *Journal of King Saud University. Computer and Information Sciences/Mağalaẗ Ğamʻaẗ Al-malīk Saud: Ùlm Al-ḥasib Wa Al-maʻlumat*, 35(8), 101651. https://doi.org/10.1016/j.jksuci.2023.101651

[20] Reddy, K. S., Rao, M. S. B., Babu, M. R., & Kumar, P. (2023). A sophisticated semantic analysis framework using an intelligent tweet data clustering and classification methodologies. *Microprocessors and Microsystems*, 98, 104793. https://doi.org/10.1016/j.micpro.2023.104793

[21] Amir, A. et al., (2013). Cognitive computing programming paradigm: A corelet language for composing networks of neurosynaptic cores, in *Proceedings of the International Joint Conference on Neural Networks*. 10.1109/IJCNN.2013.6707078

[22] Dwore, R. B. & Matarazzo, J. The behavioral sciences and health education: Disciplines with a compatible interest? *Health Education*, 12(3), 4–7. PMID: 6792131.

[23] Cuzzocrea, A. et al., (2011). Analytics over large-scale multidimensional data: The big data revolution!, in *Proceedings of the ACM 14th International Workshop on Data Warehousing and OLAP*, pp. 101–104.

[24] Bransford, J. D. (1979). *Human cognition: Learning, understanding and remembering*. Wadsworth,
[25] Gliozzo, A. et al., (2013). Semantic technologies in IBM Watson, available at www.researchgate.net/publication/270878679_Semantic_Technologies_in_IBM_Watson.

Chapter 6

Generative AI Algorithms and Techniques for Cognitive Computation

A Comprehensive Study

Smita Bharne

6.1 INTRODUCTION

Artificial intelligence (AI) has emerged as a remarkable breakthrough in the field of generative artificial intelligence (GAI). GAI has been a prominent field of study in various domains, with new advanced algorithms and profound applications in the fields of natural language processing (NLP), creative imaging art, content generation, etc. GAI algorithms produce new data with their inputs and have become an essential tool with the evolving applications in cognitive computation (or, CC, which will be referred in full going forward) (Feuerriegel, Hartmann, Janiesch & Zschech, 2023; Fui-Hoon Nah, Zheng, Cai, Siau & Chen, 2023). This chapter focuses on the theoretical underpinnings and practical approaches of the GAI algorithm to understand its potential application in various diverse domains.

GAI is different from discriminative artificial intelligence (DAI). DAI focuses on learning patterns and making predictions from the existing data. The GAI develops models and algorithms to produce AI data that closely resembles real-world data. (Nguyen, Tulabandhula & Watson-Manheim, 2023). This capability of the GAI algorithms has various implications in cognitive computing processes such as problem solving, creative art designs, voice assistance, drug discovery, and personal assistants. The entertainment, healthcare, financial, and other industries can benefit significantly from the capability to generate new and realistic data. GAI enables synthetic image creation, text output, creating music, and human-like chatbots.

The chapter begins by discussing GAI algorithms and techniques for cognitive computation, exploring fundamental concepts such as generative models, latent spaces, and probabilistic frameworks. It also focuses on the intersection of GAI with cognitive computation and discusses real-world applications in different areas. The chapter also explores current challenges, ethical considerations, and future directions or applications of GAI in cognitive computation.

DOI: 10.1201/9781003501220-6

6.2 BRIEF OVERVIEW OF GENERATIVE AI AND ITS SIGNIFICANCE IN COGNITIVE COMPUTATION

Historically, it was believed that only humans could execute creative jobs like writing poetry, inventing software, designing apparel, and composing melodies. Recent advancements in AI have transformed this presumption by generating content that is hard to differentiate from human artistry. The term generative artificial intelligence or GAI (GAI or generative AI will be used interchangeably going forward) refers to the ablity to create new content, such as text, audio, and images, based on existing training data. The popularity of wide diffusion models like DALL-E2, GPT-4, and Copilot is currently blowing up the scene. Instances such as DALL-E2, GPT-4, and Copilot demonstrate how this technology is currently changing the scope of work and communication. GAI systems can write and draw like humans and help humans answer intelligent questions. GAI can be used for information technology (IT) professional to support transitional knowledge activities, writing codes, routine demands like cooking recipes and medical advices (Feuerriegel et al., 2023). According to Goldman Sachs (2023), generative AI might increase global GDP by 7% and replace 300 million knowledge jobs.

GAI systems are taught on vast volumes of data with many training samples, unlike discriminative models that forecast current data. Cognitive computation studies human-like cognitive processes. Cognitive computation aims to study, simulate, and copy human cognition using computers. These talents include perceiving, learning, logic, decision-making, and problem-solving. Cognitive computing relies on generative AI algorithms to imitate human cognitive processes including creativity, imagination, and complicated pattern recognition.

In several sectors, GAI algorithms improve cognitive activities and applications by creating fresh data that closely resembles human-provided material. One of the important significances of GAI in cognitive computation domain has been in the field creativity and innovation. Due to the advanced form of GAI the AI-based artwork, designs, realistic images, music composition, NLP tasks including text generation, machine translation, and dialogue systems, chat, and voice assistance are possible. By learning natural language data structure and patterns, generative models (GM) may generate coherent and contextually relevant text, making human-machine interactions more natural. Virtual assistants, chatbots and language generation for creative writing and narrative benefit from this skill. Generative AI algorithms are computational systems that demonstrate creativity in language processing and excel in cognitive tasks, such as generating images and videos with unique and diverse visual content. For instance, generative adversarial networks (GANs) are the general-use medium for image generation, which makes the technology capable of imaging things diversly. Thus, now we can generate great images that look almost as good as the real ones (Karthika & Durgadevi, 2021). This ability is helpful in the perspective of computer graphics, virtual reality, and visual graphics since these fields need highly realistic visual material. GAI can contribute to the area of cognitive computing with data augmentation, and anomaly detection. The field of GAI is continuously evolving and pushing the boundaries of GAI into cognitive computation with the advent of innovation and transformations (Dipaola, Gabora, & McCaig, 2018). The rapid identification of interesting drug candidates by GAI is transforming drug discovery. AI algorithms educated on massive databases of compounds and their properties can develop unique molecules with specified functions. This allows researchers to virtually screen libraries of potential drugs, prioritize the most effective ones, and optimize existing leads for better efficacy or reduced side effects. This powerful approach holds immense promise for developing new treatments and improving healthcare outcomes (Martinelli, 2022).

6.3 IMPORTANCE OF GAI FOR VARIOUS APPLICATIONS

Due to the capability to produce new data, that resembles the input distribution, GAI algorithms are very useful from people to experts for a multitude of applications across several fields which include drug discovery, computer vision, creative design, NLP, etc. (Tripathi, Augustin, Dunlop, Sukumaran, Dheer, Zavalny, Haslam, Austin, Donchez, Tripathi & Kim, 2022). Figure 6.1 shows the GAI applications in different domains. Generative algorithms in NLP enable the production of consistent and context-applicable text, which can be used in human-machine dialog systems, document translation, text summarization, and questions and responses.

The computer vision field benefits from these algorithms by creating lifelike pictures, as well as videos, applicable in the field such as image synthesis, style transfer, and virtual reality. The fact that generative algorithms provide various solutions to creative design as they can be used in art-making, music, and literature, with a vast range of variations that humans may not have seen before. For the progress of novel drugs and molecules, the algorithms are equally important as they help to generate new structures with specific properties, thus reducing drug development time and enhancing the discovery of new therapeutic agents. Proper knowledge of generative algorithms is essential for developing their potential for various applications and encouraging innovation and development in electricity and power generation (Dipaola et al., 2018). Table 6.1 shows some of the potential application areas where GAI can play an important role (Fui-Hoon Nah et al., 2023; Nguyen et al., 2023)

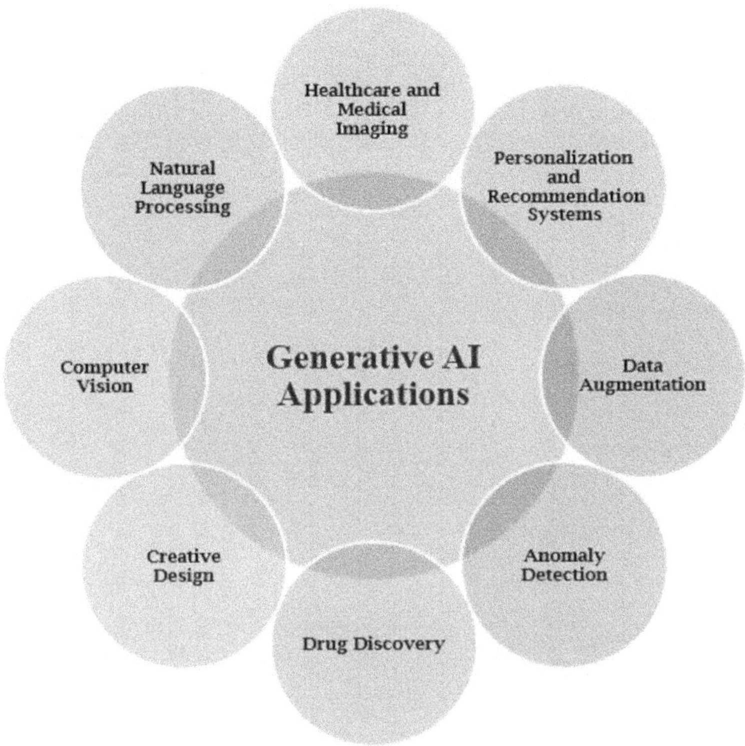

Figure 6.1 Generative AI Applications in Different Domains.

6.4 OBJECTIVE OF THE CHAPTER

The objective of the chapter is to focus on how generative AI algorithms are relevant in the domain of cognitive computing. The chapter gives an overview of both theoretical foundations and practical applications of generative AI algorithms in a manner that enables readers to recognize their relevance to the emulation of human cognitive processes and the advancement of cognitive capabilities within AI systems.

6.5 FUNDAMENTALS OF GENERATIVE AI

GAI is a revolutionary concept in machine learning (ML) because it can generate new data similar to its well-trained input data. A solid understanding of the fundamentals of GAI is needed in order to understand its theories, processes, and applications in numerous fields. This section highlights the concepts of GAI and how it differs from discriminative AI.

Table 6.1 Importance of GAI with Different Application Areas

Application	*Importance of Generative Algorithms*	*Citations*
Natural language Processing	GAI make coherent, contextually appropriate text for dialogue systems, language translation, and text summarization.	(Vaswani et al., 2017)
Computer vision	The task of GAI is of utmost importance since it is responsible for the production of life-like images and videos, and its uses include style transfer and virtual reality applications.	(Dosovitskiy et al., 2021)
Creative design	GAI is useful in creative design as it helps generate artwork, music, and literature, enhancing human creativity.	(Pumarola et al., 2018)
Drug discovery	GAI systems are used in drug discovery, have a central hand in producing new molecular compounds of chemical compounds with some characteristic properties, thus, the process of drug discovery is faster.	(Gangwal et al., 2024)
Anomaly detection	GAI algorithms understand the normal data distribution and recognize outliers in diverse domains to discover anomalies.	(G. Liu et al., 2022)
Data augmentation	GAI algorithms improve machine learning model resilience and generalization in numerous applications by adding synthetic data to restricted datasets.	(Antoniou et al., 2017)
Personalization and recommendation systems	GAI algorithms help personalization and recommendation systems by creating individualized content or recommendations based on individual users' interests.	(Y. Zhang et al., 2024)
Healthcare and medical Imaging	Healthcare and medical imaging are increasingly using generative AI algorithms for image synthesis, disease prediction, and treatment planning.	(Kuzlu et al., 2023)

6.6 FUNDAMENTALS OF GAI AND ITS DISTINCTION FROM DISCRIMINATIVE AI

Generative AI is core to the idea of generative models which use statistical models for probability distribution of input data. These models aim at grasping information regarding the joint probability at the feature labels. There are two types of generative models: (1) explicit ones and (2) implicit ones. In explicit generative models, the parameters are directly determined by the model. However, implicit generative models can generate data points without specifying probability distributions (Meila, 2006).

6.7 KEY TERMINOLOGIES IN GAI

Expanding on the previous information, GAI models are created using a ML-based transformer architecture, such as a deep neural network. This type of model can generate new data samples by leveraging the patterns it has learnt. The key concepts of GAI are discussed in the below section.

1. **Large language model (LLM):** AI is the scientific study of designing machines capable of performing functions that typically require human intelligence. Natural language processing teaches machines to understand, interpret, and generate human language. A language model (LM), on the other hand, can refer to any of these two things: (1) a computational statistical model for word count predictions in sentences or (2) a whole document, depending on how it is used by its creator(s). For instance, when writers discuss language models, they could be talking about computer programs designed with algorithms mimicking human speech patterns. The most common type of such a model is recurrent neural networks (RNNs). An RNN is a type of artificial neural network that allows for two-way interactions between nodes within the same layer and across different layers since they contain internal feedback loops just like how long-term memory functions within our brains which stores information for future reference. Examples of prominent big language models are BERT and GPT-3. BERT training comprises of 340 million parameters while GPT-3 comprises of 175 billion parameters (Devlin, Chang, Lee & Toutanova, 2019).
2. **Reinforcement learning from human feedback (RLHF):** RLHF learns from sequential tasks (series of chats) based on humanize feedback. RLHF is training a reward-based model on humanized feedback. This model is then used to optimize the policy using data-efficient and resilient algorithms (Bakker, Chadwick, Sheahan, Tessler, Campbell-Gillingham, Balaguer, McAleese, Glaese, Aslanides, Botvinick & Summerfield, 2022). The application example of RLHF

is mainly used in conversational model like ChatGPT, Sparrow, InstructGPT (Kucharavy, Plancherel, Mulder, Mermoud & Lenders, 2024); Ouyang, Wu, Jiang, Almeida, Wainwright, Mishkin, Zhang, Agarwal, Slama, Ray, Schulman, Hilton, Kelton, Miller, Simens, Askell, Welinder, Christiano, Leike & Loweet, 2022) to generate chat messages that follow established human preferences for length, style, and appropriateness. Fluency plays into RLHF's successful track record in some ways. Its main challenge is insufficient public datasets or resources for coding, which make people assume that it is an NLP for experts only.

3. **Prompt learning:** One approach to LLMs is prompt learning, which takes advantage of the information already present in language models to perform subsequent tasks (P. Liu, Yuan, Fu, Jiang, Hayashi & Neubig, 2023). Prompt learning is usually effective and adaptable without fine-tuning the language model. Whenever any instruction is given in the form of a prompt, the sentiment can be shown as either positive or negative. The intricate prompt engineering, is now possible due to recent advancements (P. Liu et al., 2023). While giving prompt instructions it may be possible that the model may need more information to understand the context. The response quality not only depends on prompts but also on how the prompts are continuously updated to maintain their relevance about new questions from the customers and their feedback for continuous improvement. One such example in a real-world scenario is providing correct and useful responses to the common questions from customer care operations services in a company. In this situation, any company can design specific prompts to generate relevant responses.
4. **Sequence-to-sequence model (seq2seq):** ML methods map input to output sequences. ML is used to translate languages. These seq2seq methods have two parts: (1) each word in a text is encoded into a hidden vector with its context by encoders. In the other direction, (2) the decoder converts the vector into a word in the new language, taking into consideration linguistic dependencies from the previous output. Text-to-image and text-to-speech mappings are now possible with seq2seq models. (Yousuf, Lahzi, Salloum & Shaalan, 2021).
5. **Transformer model:** It is a deep learning (DL) model known as the Transformer which is superior to RNNs and convolutional neural networks (CNNs) in terms of processing sequential data. The Transformer architecture's main feature is self-attention (Han, Xiao, Wu, Guo, Xu & Wang, 2021). This feature enables the model to decide day-by-day weights for various inputs. Thus, it produces scores that represent the extent to which an individual input element is important relative to others, thus facilitating much more efficient encoding of longer dependencies and contextual relationships than

traditional RNNs. Different from RNNs which process sequences one step at a time, Transformers can process an entire sequence at once due to their parallelizable nature. This makes it possible to apply parallel processing that greatly expedites computational efficiency and scalability aspects since the computation is not so dependent on the order of elements present in memory at any given point of time within sequences processed by these models. Machine translation as well as text summarization are some of the tasks where its usage becomes particularly privileged such as sentence completion problems in which it is going to be necessary to go far beyond the nearest context of a word for correct completion.

6. **Zero-shot learning:** Zero-shot learning is a model in ML to tackle the issue of scarcity of acquisition data. One of them is when you teach a machine without giving it direct access to any data. On the other hand, the former views learning from just a few examples as possible scenarios. The reason why these two types of learning are needed in practice is because they help reduce costs related to AI systems' implementation. For instance, LLMs can learn very fast with a few examples like predicting sentiment in reviews. (Xian, Schiele & Akata, 2017). This adaptability leaves LLMs as versatile tools serving many different purposes and favorite amongst the users.

6.8 DISTINCTION FROM DISCRIMINATIVE AI

The objective, methodology, and applications of generative AI and discriminative AI are entirely different. Discriminative AI models like logistic regression, support vector machines (SVM), and deep learning determine the decision boundary or hyperplane that separates different groups or classes in the input data (Meila, 2006).Generative AI systems are about learning and creating new examples from the whole probability distribution of the input data. Discriminative AI techniques are commonly employed in classification problems including regression analysis, and pattern recognition tasks whereas generative AI methods can be used for instance when synthesizing images reconstructing images assimilating texts creating fake versions of such things, etc. (Feuerriegel et al., 2023)

6.9 KEY COMPONENTS OF GENERATIVE MODELS: PROBABILISTIC FRAMEWORKS AND LATENT SPACE

In ML, generative models obtain fresh data samples with similar features to the input data distribution. The focus of this section is on some of the major ideas involved in generative techniques including latent space and probabilistic frameworks. This chapter discusses two critical aspects: (1) probabilistic frameworks, and (2) latent space representations.

6.9.1 Probabilistic Frameworks

Probabilistic representations and probabilistic models make use of chance to fix mathematical models for complex data. The frameworks utilize probability to perform calculations or to make decisions based on a set of derived likelihoods. Uncertainty that is most often referred to here as probabilities in classical statistics. A variational autoencoder (VAE) takes into account these methodologies to comprehend how data works out for its various components. This is done by, sampling from latent space distribution which helps it to capture patterns within input space. (Harshvardhan, Gourisaria, Pandey & Rautaray , 2020).

6.9.2 Latent Space Representations

AI scientists in the world have made significant advancements, especially in the field of AI programming languages and semantics. This has been brought about by using tensors to represent a given piece of data on one hand while on the other hand applying deep neural networks that allow automatic translation into higher-dimensional spaces, where these two main elements are intertwined through their common property-space mapping framework or notation system (Gelada, Kumar, Buckman, Nachum & Bellemare, 2019). Table 6.2 shows the tabular analysis of key components of GAI along with applications, merits, and demerits with recent advancements.

6.10 GENERATIVE LEARNING MODEL AND TECHNIQUES

The generative learning model is a type of ML model that creates new data that look a lot like training data. Graphics processing units (GPUs) are needed to train large-scale generative models on massive datasets. The different types of generative models are illustrated in Figure 6.2. In 2014 (Goodfellow, Pouget-Abadie, Mirza, Xu, Warde-Farley, Ozair, Courville & Bengio, 2014), introduced GANs that comprise two neural networks; one network produces fake images and the second one detects true or fake images. VAEs proposed by Kingma and Welling (2019), produce new visuals by compressing inputs. Flow-based models introduced by Rezende and Mohamed (2015) reverse the probability distribution of input data. Recent research by Ho, Chen, Srinivas, Duan & Abbeel (2019), describes diffusion models, as a significant class of generative models because they can create high-quality examples from complex datasets.

Table 6.2 Analysis of Key Components of GAI

Component	*Methodology*	*Applications*	*Advantages*	*Limitations*	*Recent Developments*	*Citations*
Generative models	Autoencoders, GANs, Flow-Based Models, Bayesian Gen	Image Generation, Text Generation, Anomaly Detection	Capture complex data distributions	Mode Collapse in GANs	Improved stability and diversity in GAN training	(Goodfellow et al., 2014)
Latent spaces	Learned representations of data	Image Generation, Dimensionality Reduction	Compact and informative representations	Disentanglement and interpretability challenges	Disentanglement techniques in VAEs and GANs	(Kingma & Welling, 2019)
Probabilistic frameworks	Variational Inference, Bayesian Inference	Uncertainty Estimation, Probabilistic Modelling	Probabilistic uncertainty estimation	Variational approximation errors	Advances in amortized variational inference	(Rezende & Mohamed, 2015)

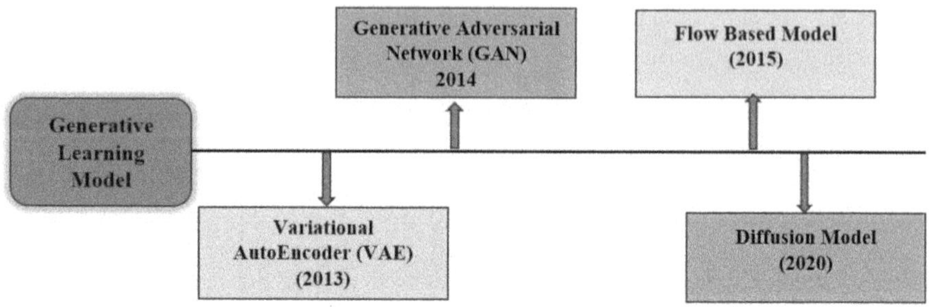

Figure 6.2 Different Forms of Generative Learning Model.

6.11 OVERVIEW OF DIFFERENT TYPES OF GENERATIVE ALGORITHMS AND THEIR UNDERLYING PRINCIPLES

Generative algorithms refer to diverse methods used to find the actual data distribution as well as generate new samples that mimic the training set. The ensuing section provides an in-depth explanation of various kinds of generative algorithms. Regarding the generation of data, all of them come with their unique principles and benefits leading to a wide range of generative algorithms which are found in the ML domain, which is discussed in this section.

6.12 VARIATIONAL AUTOENCODERS

Variational autoencoders (VAEs) are often used for ML (which, as a reminder to the reader, stands for machine learning) without supervision and generating new data samples. VAEs are different from normal autoencoders in that they study probabilistic mappings between input data and output data. Thus, they can catch the hidden distribution in input data by introducing latent variables.

The representation learned through the VAE technique should capture the latent information in the input data that is essential for its accurate reconstruction. This section fundamentally introduces VAEs, explaining what they are built on and their applications in life (Kingma & Welling, 2019).

VAEs are essentially probabilistic generative models that comprise an encoder network, a decoder network, and a latent variable distribution. Let x represent an input data sample, z represent the latent variable, and θ denote the parameters of the encoder and decoder networks. Encoding network $q(z|x)$ approximates the true posterior distribution $p(z|x)$ through mapping input data x to a distribution over latent variables z. A decoder network $p(x|z)$ samples a point from the latent space z to reconstruct input data x. During training, the goal is to optimize parameters θ such that the evidence lower bound (ELBO) is maximized:

$$ELBO(\theta) = Eq\theta(z \mid x)\left[logp\theta(x \mid z)\right] - KL\left[q\theta(z \mid x) \parallel p(z)\right] \quad (1)$$

where the first term represents the reconstruction loss while the second term corresponds to the KL-divergence between approximate posterior $q\theta(z|x)$ and the prior distribution $p(z)$. The ELBO encourages meaningful latent representations and low reconstruction error when maximized (Rezende & Mohamed, 2015).

In practice, VAEs optimize the ELBO objective during training using stochastic gradient descent (SGD) or its modifications. Perr mini-batch of input data samples, an encoder network computes parameters for the approximate posterior distribution $q(z|x)$ whereas the decoder network generates reconstructed samples in the latent space. After that, Monte Carlo estimation is applied to solve the ELBO, and the so-obtained gradients are used for backpropagation through the network updating θ. By doing this iteratively until it converges, a VAE model is achieved hence capable of generating new samples from this space which have been learnt.

6.13 GENERATIVE ADVERSARIAL NETWORKS

In 2014, Ian Goodfellow and colleagues devised generative adversarial networks (GANs), a deep learning framework. The generator and discriminator neural networks are trained in simultaneous competitions to establish a competitive adaptation. Figure 6.3 shows the architecture of GAN

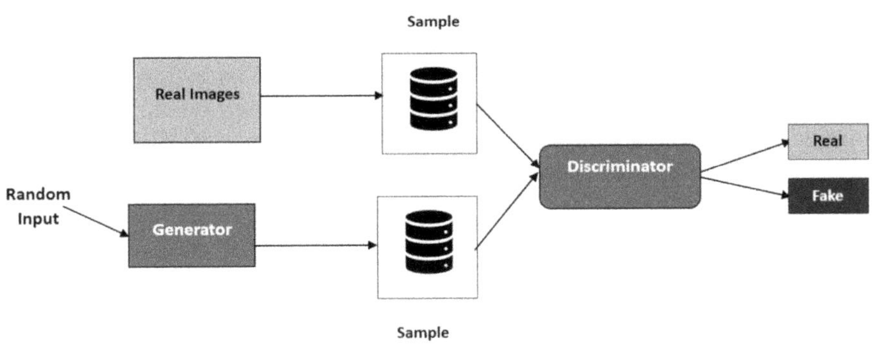

Figure 6.3 Architecture of Generative Learning Models.

(Nelson, 2020; Gui, Sun, Wen, Tao, & Ye, 2023). The generator is taught to generate reality-like data samples (images, text, etc.) from random noise or latent representations, while the discriminator learns to distinguish artificial data from both training and generator data. At a higher conceptual level models in GANs use notions from game theory whereby each time these two players try to achieve their objectives iteratively by playing a minimax game. The generator seeks to produce outputs that closely match those from actual data, on the other hand discriminator seeks to label correctly between actual and generated samples. At the system level, GANs require optimizing training methods while managing convergence concerns and maintaining a balance between generator and discriminator operations. Strategies including batch normalization, learning rate scheduling, or adjustments in architecture such as deep convolutional GANs play an important role here. The framework of Generative Adversarial Networks is handy for generative modelling. (Gui et al., 2023).

The main concept of GANs is training a network (x) to generate samples that resemble those in real-life scenarios. On the other hand, they involve another network (z) trained such that it discriminates between genuinely observed samples from those created artificially. We define z as a random noise vector taken from prior distribution $p(z)$, $G(z)$ as output produced by the generator network when input with noise z and x as data sample. Discriminator (x) will hence give out a number showing how likely it is that x has been generated through G instead of belonging to the true dataset.

$$minGmaxDEx \sim pdata(x)\left[logD(x)\right] + Ez \sim pz(z) \left[log\left(1 - D\left(G(z)\right)\right)\right] \tag{2}$$

The discriminator aims to maximize the probability of correctly classifying real and fake samples, while the generator aims to minimize this probability by generating realistic samples that can fool the discriminator. The discriminator and generator networks are adversarial trained iteratively. The discriminator is trained to maximize the probability of assigning high scores to real samples and low scores to created samples given a mini-batch of real data samples. Meanwhile, the generator is trained to reduce the discriminator's likelihood of accurately categorizing created samples as fake.

6.13.1 Flow-based Generative Model

Data distributions, flow-based generative models represent a particular type of probabilistic model that learns invertible transformations unlike those traditional GANs and VAEs (Bond-Taylor, Leach, Long & Willcocks, 2022). Unlike GANs or VAEs which generate samples directly from latent space, flow- models transform simple samples (e.g., from a Gaussian distribution) into more complex target distributions (e.g., the distribution of training data) during training (Ho et al., 2019). Trained on optimizing likelihood for transformations learned by observing data, they enable efficient sampling and density estimation at a time using them to perform inference on observed data. There have been several applications where these flow-based models have shown good performance including image generation, density estimation, and data compression among others though they have different shortcomings including computational complexity associated with training them and inability to capture long-range dependencies within data distributions.

Flow-based models have a lot to offer when it comes to exact likelihood evaluation and sample efficiency since both forward and inverse transformations can readily be computed. This makes them suitable for tasks that require accurate modeling of data distribution, like density estimation.

6.13.2 Diffusion Model

Due to synthesis high-quality samples from complex data distributions, generative diffusion models have been popular recently. These are models that use diffusion processes to represent the distribution of data where noise is successively added at every step to a starting point in the sample. The generative diffusion models work, by aggregating denoised data samples. From a noisy input that represents any arbitrary member in the real distribution population or community, a stepwise reduction of the noise leads to a sample very close to one individual representing whole lots of them. At each stage, neural networks are applied for denoising purposes which behaves as a layer (Yang, Zhang, Song, Hong, Xu, Zhao, Zhang, Cui & Yang 2023)

Image production, audio synthesis, and natural language processing benefit from generative diffusion models. For example, diffusion models have been employed for generating realistic images that possess great precision during image synthesis. This in turn allows the acquisition of detailed photorealistic images while filtering out unnecessary noise generation scenarios considering some parameters specific to the domain being addressed through generation algorithms (Yang et al., 2023). In audio domain generation tasks, these methods guarantee that individuals have free control over the sounds produced by the model. If a person were to speak fluently without referring to any written text, their speech would likely remain coherent, guided by the patterns and motivation derived from previously learned sounds during training. This can be used in conversations, and stage plays to simulate dialogues as well as narrations.

Table 6.3 shows the comparative analysis of the prominent generative modeling techniques concerning the objective function, generation quality, training stability, complexity, and other parameters.

Table 6.3 Comparative Analysis of Various Generative Modeling Techniques

Parameter	*Autoencoders with variants*	*VAEs with probabilistic modelling*	*GANs*	*Flow-based models, Diffusion models*
Description	Decode and recreate input data from latent space.	Learn a data-representation probabilistic distribution in latent space.	Train a generator to generate real-looking data samples while a discriminator learns to distinguish them.	Use invertible transformations to convert simple to complicated target distributions.
Latent Space Representation	Continuous, typically lower-dimensional space.	Continuous, probabilistic distribution (e.g., Gaussian).	None explicitly defined, implicitly learned by the discriminator.	Continuous, invertible transformation of the input space.
Training Approach	Supervised or unsupervised learning using backpropagation.	Optimization of evidence lower bound (ELBO) using stochastic gradient descent.	Minimax game between generator and discriminator networks using alternating optimization.	Optimization of likelihood under the learned transformation using gradient-based methods.
Objective Function	Reconstruction loss or combination of reconstruction and regularization terms.	ELBO objective, combining reconstruction error and Kullback-leibler (KL) divergence.	Binary cross-entropy loss between actual and produced samples.	Data log-likelihood is negative under the learned distribution.
Generation Quality	Limited by the reconstruction ability of the decoder network.	Moderate, with controlled randomness in sampling from the latent space.	High, often surpassing other methods in visual fidelity.	High, with fine-grained control over sample quality and diversity.
Training Stability	Sensitive to initialization and hyperparameters.	Sensitive to KL annealing and choice of prior distribution.	Prone to mode collapse and training instability.	Stable training due to invertibility and explicit likelihood modelling.
Citations	(Goodfellow et al., 2014; Kingma & Welling, 2019)	(Rezende & Mohamed, 2015)	(Goodfellow et al., 2014)	(Kingma & Welling, 2019; Tschannen et al., 2020)

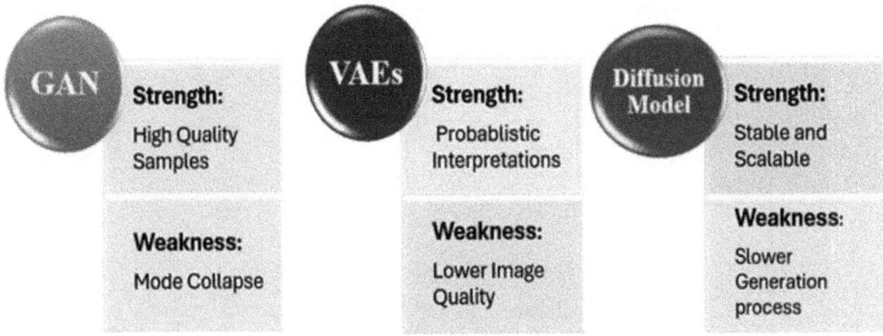

Figure 6.4 Major Strength and Weakness of GANs vs. VAEs Diffusion Models.

6.13.3 Strengths and Weaknesses of each Technique

Generative models have advanced rapidly, bringing both notable strengths and inherent weaknesses. We present a comparative analysis of the strengths and weaknesses of VAEs, GANs, flow-based models, and diffusion models. Figure 6.4 shows the major strength and weakness of GANs versus VAEs diffusion models.

Researchers can determine which model is optimal for a certain application by analyzing the strengths and limitations of each strategy. Table 6.4 shows a comparative analysis amongst the various generative modeling techniques.

6.14 APPLICATIONS OF GENERATIVE AI IN COGNITIVE COMPUTATION

Innovations like creative design and art, image generation and manipulation as well as dialogue systems that utilize natural languages could be built on AI-based on GAIs. Generative AI offers innovative solutions that simulate human cognitive functions, thereby enhancing how humans comprehend, reason, and express themselves through ideas or creative works.

6.15 OVERVIEW OF COGNITIVE COMPUTATION AND ITS INTERSECTION WITH GENERATIVE AI

This section explores the concept of cognitive computation (CC) and examines how generative AI supports and enhances cognitive tasks. Cognitive computation studies the computational principles behind human thinking to come up with mimetic or enhancement model algorithms. The intersection of cognitive computation and generative AI enhances human-like creativity, imagination, and complex problem-solving.

Table 6.4 Comparative Analysis of Strengths and Weakness of the various Generative Modeling Techniques

Technique	*Strengths*	*Weaknesses*	*Citations*
Variational autoencoders (VAEs)	• Probabilistic framework with uncertainty estimates. • Continuous latent space allows for smooth interpolation. • Good regularization due to Kullback-leibler (KL) divergence.	• Often produces blurry outputs. • Balancing reconstruction loss and divergence is challenging. • Approximation of the true posterior may not be perfect.	(Kingma & Welling, 2019; Rezende & Mohamed, 2015)
Generative adversarial networks (GANs)	• Can generate highly realistic and sharp images. • Flexible architecture applicable to various data types. • Strong performance in image synthesis and style transfer.	• Training instability due to its adversarial nature. • Mode collapse: limited sample diversity produced by generator. • Requires extensive hyperparameter tuning.	(Goodfellow et al., 2014; Xian et al., 2017)
Flow-based models	• Exact log-likelihood computation possible. • Invertible transformations allow for easy sampling and density estimation. • Strong performance in tasks requiring exact data distribution modeling.	• High computational cost due to invertibility constraints. • Complex design with many parameters can impede training compared to simpler models.	(Asim et al., 2020; Tschannen et al., 2020)
Diffusion models	• Robust training process with less mode collapse. • Can generate high-quality and diverse samples. • Suitable for tasks requiring gradual refinement of samples.	• Computationally intensive due to iterative denoising steps. • Slower generation process compared to other models. • Requires careful tuning of the diffusion process parameters.	(Brown et al., 2020; Grathwohl et al., 2020)

6.15.1 Overview of Cognitive Computation

The concept of cognitive computation emphasizes understanding the computational processes underlying human cognition from an interdisciplinary perspective. It tries to answer questions about how we see things, think, learn, and choose to create machines modelling these processes. It combines knowledge gained in cognitive psychology, and neuroscience as well as linguistics philosophy, and artificial intelligence. With computational techniques and theories, researchers want to develop intelligent systems that can do things humans normally do through reasoning like natural language understanding, problem-solving, or decision-making.

CC does not follow traditional rule-based approaches used in AI development. While traditional AI is based on pre-set rules and logic, this cognitive science approach focuses on learning through data and past experiences instead. Thanks to ML algorithms, it lets the machines learn as they go along making them better each time at handling complex problems while remaining relevant in various fields it is applied at any given moment. It enables the perception and reasoning processes of the AI system to operate like that of humans. These include features such as understanding human language, perceiving information from senses or even making decisions that are significant to the context. It is through deep learning techniques and neural networks that CC models can generate useful knowledge from unstructured data such as allowing them to behave more humanly in conversation responses to the user (Gutierrez-Garcia & López-Neri, 2015)

6.15.2 Intersection of Cognitive Computation with Generative AI

CC and GAI intersection is one of the most interesting areas of AI research. For instance, cognitive computing is inspired by the way humans think and includes processes such as making sense of information, being logical about things, making decisions as well as identifying new patterns. On the other hand, these two domains synergize from neural networks to complex probability levels, and it helps them understand that what they get as input becomes more creative than ever before for intelligent responses. Purposefully composing original texts supporting them not only intellectually but also emotionally creates a humane form of communication. Furthermore, generative AI algorithms are supposed to be used in cognitive computing structure aiming at establishing more human like functionalities conveying feelings artificial intelligence and cognitive science at large. Figure 6.5 shows the intersection of cognitive computation process with generative AI algorithms. These generative models, in turn, are applied in various cognitive tasks and applications. The detailed applications of GAI in the CC domain are explained in section 4.2.

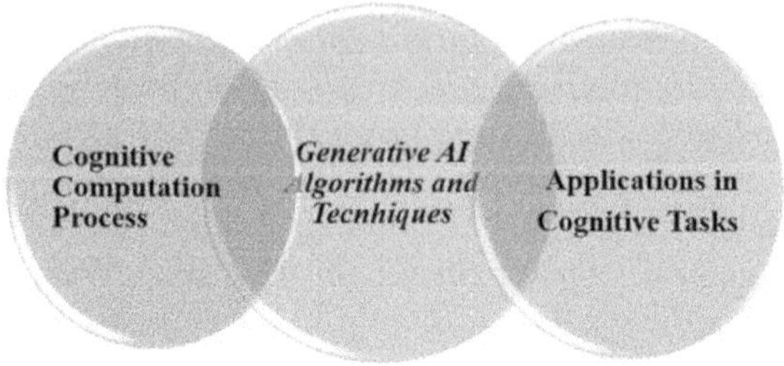

Figure 6.5 Intersection of Cognitive Computation with Generative AI.

Following some of the applications that intersect in the domain of CC with GAI (Aghav-Palwe & Gunjal, 2021; Chamola, Sai, Sai, Hussain & Sikdar, 2024; Malloy & Gonzalez, 2024). CC draws inspiration from human cognition to design and develop intelligent systems. By understanding how humans perceive, reason, and learn, researchers can design generative models that mimic human-like behavior and generate data samples that are perceptually and semantically meaningful. Cognitive computation aims to improve human-computer interaction by developing systems that understand and respond to human inputs in natural and intuitive ways.

1. **Improving human-computer interaction:** To improve user experience in general, GAI can help by providing natural language responses, realistic visual outputs, and individualized recommendations.
2. **Simulation of human creativity:** Being creative and imaginative in thought as part of human cognitive activities enables us to come up with fresh ideas, pictures, or even insights. According to Forbes (2020), generative AI technologies including "creative AI" as well as "style transfer" seek to reproduce human creativity by producing numerous different outputs depending on learned patterns and styles such as creative AI.
3. **Personalized recommendation systems:** A generative personalized recommendation system using CC will enable businesses to generate customized suggestions about products, services, or content from consumers' behavior and preferences while still considering them as individuals. Generative inferring allows for such suggestions based on experience data from their profiles and past interactions.
4. **Natural language understanding and analysis:** In the sphere of cognitive computing, generative AI has found an important convergence with natural language processing abilities such as chatbot conversation generation, sentiment analysis, and document translations. Principles employed in human cognitive science for understanding, decoding languages through which they communicate with each

other in various media, generative techniques may be used in making relevant coherent answers to different questions placed in a given context.

6.16 REAL-WORLD APPLICATION OF GAI IN COGNITIVE TASKS

There are many areas where GAN techniques have been applied to cognitive tasks. We give an intricate view of a few practical uses of generative AI in cognitive tasks.

6.16.1 Text Generation in NLP

Text generation tasks play a vital role in AI; these include language translations, dialogue generation and content creation etc. (Pokhrel & Banjade, 2023). Some of the current successes of this technology allows include coherent and contextually relevant text generation in many languages and domains by transformer models such as GPT (Stiennon, Ouyang, Wu, Ziegler, Lowe, Voss, Radford, Amodei & Christiano, 2020)as well as Bert (Devlin et al., 2019). RNNs, transformers and language models like GPTs can help generate summative conversation translations among others. Some machine translation services, and virtual assistants use language models like ChatGPT to make authoring news articles, product descriptions, social media postings, and answering queries easier.

These research models based on large language processing are being used worldwide including inbuilt applications (Bakker et al., 2022; Vaswani, Shazeer, Parmar, Uszkoreit, Jones, Gomez, Kaiser & Polosukhin, 2017).

Conversational AI has gained a lot of attention as a chatbot-type system that includes LLMs like GPT-3, formatting models such as the Palm talk tool or Galactica model to ask us anything model such as Llama-2 or Llama-3(Large Language Model Meta AI,). It is an autoregressive language model based on a transformer architecture (Raiaan, Mukta, Fatema, Fahad, Sakib, Mim, Ahmad, Ali & Azam, 2024). The aim is to make these chat-conversations, robots able to carry out different tasks like demand forecasting; inventory optimization beneficial to society. The ongoing research on LLMs depicts a future full of promise for conversational AI (Goertzel, 2023; Stiennon et al., 2020)

6.17 IMAGE GENERATION AND MANIPULATION FOR CREATIVE DESIGN AND ART

Art and creative design have seen a revolution thanks to AI, nowadays as a backbone to all processes of image creation and manipulation. Image generation in AI domain is rapidly growing, since DALL-E2 was introduced in 2022. It is popular since it can produce images based on prompts of texts or changing already existing ones. The field of image-generative AI has witnessed tremendous growth thanks to the introduction of DALL-E2 in 2022(Alhabeeb & Al-Shargabi, 2024). This technique is very useful in both artistic and professional contexts since it can help create visual content automatically using words provided and alter ready-made ones. For example, there are GANs, and VAEs that help to create high-resolution images, such as manipulating visual attributes of an image (e.g., style transfer, image-to-image translation) or restoring a missing part from an image (e.g., inpainting).

The roles played by image generation and manipulation have always been critical in various fields, expanding cognitive capabilities and introducing novel applications. Within the medical domain, using GAI methods to generate synthetic medical pictures is one example. With the help of diverse and realistic medical images generated by GAI techniques, researchers can enhance their datasets thus improving machine learning models' performance for disease diagnostic tasks like medical image analysis as well as for training purposes in medical imaging where synthetic medical images are required.

The use of GAI algorithms that perform generative functions also facilitates image-based communication interactions within cognitive tasks like human-computer interfaces (HCI) and augmented reality (AR). For example, these algorithms allow for the creation of avatar or virtual reality environments that appear authentic thereby creating deeply engaging experiences within such applications as VR gaming systems. Moreover, in educational environments, these tools which are powered by AI technology apply to producing interactive visual materials thereby assisting learners to understand complicated subjects quite well during practice sessions paired with multimedia presentations at classes or other venues where they study together with their tutors. Creative design and content creation utilize image generation and manipulation with GAI tools like painting styles, visual concept generations, and monotonous duties automatization. This is made possible by using AI algorithms in GANs that help these people streamline their work, try out different design concepts, and also produce beautiful content across various forms ranging from digital arts to advertisements among other uses(Zhou & Lee, 2024).

Several ground-breaking studies have shown how Generative AI has the potential for cognitive tasks such as image generation and manipulation. For instance,(Karras, Laine & Aila, 2019) work introduced StyleGAN, an advanced GAN architecture that can produce high-resolution pictures with rich details most close to reality ever seen. Additionally BigGAN by (Brock, Donahue & Simonyan, 2019) is a cutting-edge GAN model that produces realistic images across diverse categories and settings.

6.17.1 Music and Audio Generation

Music and audio generation have been implemented with AI techniques by creating original compositions, soundtracks, and audio effect opportunities. GAI has revolutionized music creation by allowing artists to come up with musical ideas from scratch or regular music samples. RNN, GAN are used to create music in different styles and genres like classical, jazz, electronic, and pop music (Yang et al., 2023). Such models can record music composition patterns and structures and create new songs as if performed by a genre representative or artist. In addition, generative models facilitate realistic sound effects creation, voice synthesis, and manipulation techniques among the vast possibilities used for digital audio signal processing (Engel, Agrawal, Chen, Gulrajani, Donahue & Roberts, 2019). There have been various applications of generative models in audio synthesis like Aiva, ERNIE-music, Harmonai, Infinite Album, Jukebox, Mubert, Musico, Noise2Music, Sonify, Splash AI Beatbot, Soundful, etc., which use natural language processing techniques to interpret text prompts to create music according to this text. A good example of such software is Musico, which uses additional inputs such as gesture input data, posture data, source code, or other sounds to enable more personalized music generation process (Kaliakatsos-Papakostas, Floros & Vrahatis, 2020).Generative models led by cognitive music perception and cognition could create artistically pleasing and engaging music/audio intended for human listeners according to users' music preferences.

6.17.2 Brain Network Computing

GAI application for brain network computation can be of benefit in cognitive activities. It involves simulating and modeling intricate brain connections, something that is closely related to cognitive functions like thinking, learning, remembering as well as choosing. Using GAI techniques, one wants not just to understand but to recreate neural networks' activity in the brain, this being a prerequisite for cognitive abilities. This makes it possible to produce artificial brain data which helps in enhancing actual data sets, perfecting computational models, and making research on the brain more profound. GAI algorithms like deep learning allow us to scrutinize brain images for patterns, and anomalies to help with diagnosing illnesses in neurology also. In real world, GAI can use brain models, as an effective way of communication with dumb people who have no other means of communication (Gong, Jing, Chen, Pun, Huang, Saha, Nieuwoudt, Hu & Wang, 2023)

New technologies are showing some promise in this domain. Remarkably, Meta AI's Speech from Brain and Non-Invasive Brain Recordings projects are working on decoding speech from non-invasive brain recordings. Stable

Diffusion model for Brain Images which uses diffusion models to reconstruct images from human brain activity (L. Zhang et al., 2023). These developments possess the potential to significantly enhance communication possibilities for speech-disabled individuals. The convergence between GAI within brain network computing offers hope for furthering neuroscience research while enhancing clinical outcomes within neurology.

6.18 CHALLENGES AND FUTURE DIRECTIONS

6.18.1 Current Challenges in Generative AI for Cognitive Computation

AI has made some spectacular advances in recent years; however, it is still faced with numerous barriers in cognitive computation such as broad adoption and effectiveness. Here we explore some of these challenges.

1. **Generated sample diversity:** The issue of guaranteeing that generative models produce samples that are both of high quality and exhibit diversity remains unresolved. The issues of collapse in modes and degeneracy of samples arise when these models fail to reflect the full complexity and variability available in the training data. (Fui-Hoon Nah et al., 2023)
2. **Training instability:** The training of generative models especially for complex architectures like (GANs tends to be unstable leading to issues of non-convergence. Mode collapse i.e., where few types of samples are generated and training divergence is often encountered (Cao, Li, Liu, Yan, Dai, Yu & Sun, 2023)
3. **Ethical implications and misuse:** The emergence of generative AI Technology raises moral issues relating to misuse by creating deepfakes or fake news. Responsible and ethical generative AI research and applications require well-managed and ethical considerations. (Cao et al., 2023)
4. **Bias and fairness:** AI models formulated on faulty data may sustain or enhance societal prejudices existing within the simulated world. To have fair AI systems characterized by justice and inclusion, we must remove partiality from training records besides making sure all produced results are fair (Fui-Hoon Nah et al., 2023)
5. **Scalability and efficiency:** The challenge of scalability and efficiency comes with large-scale data sets, as well as models that require a lot of computation. Real-world deployment requires optimization for computational intensity in both training algorithms & infrastructure used to process them. (Brown et al., 2020)
6. **Interpretability and explainability:** The intricacies of generative models often make it hard for users to interpret them when they

want to know why the model arrived at a certain decision or result. Interpretability determines trust and understanding in generative AI (Lipton, 2018)

7. **Collaboration between humans and AI:** However, generative AI systems do not work well with people who are engaged in the process of thinking which entails problems including making choices. Nonetheless, integrating human expertise and feedback in the domain of GAI systems is quite intricate (Fui-Hoon Nah et al., 2023)

6.18.2 Ethical considerations and societal impacts of GAI

1. **Privacy and data security:** There may be instances when GAI models trained using large data sets inadvertently capture personal details or sensitive information thereby raising questions about privacy violations and data protection breaches. When GAI systems compromise individual data, people start doubting these companies' ability to guarantee their privacy rights. To counteract these risks robust privacy-preserving approaches and security measures must be taken. In this respect, generative AI may leak confidential information (Mittelstadt, Allo, Taddeo, Wachter & Floridi, 2016).
2. **Misinformation and over-reliance:** Perhaps even more worrying is that generative AI technologies for instance deepfakes have made it possible for fake information to be created and spread across different societies posing a threat to the public discourse, political stability as well as individual reputation. Misuse of AI for creating the content and avoiding plagiarism with the AI-generated texts. Generative AI has the potential to be used for dishonesty during examinations or assignments. It is observed that GAI may generate violent, profane, or erotic content (Xu, Fan & Kankanhalli, 2023) .
3. **Legal and regulatory challenges:** GAI is advancing much faster than the legal and regulatory systems that are supposed to govern its operations. Making laws and acts that govern AIs such as liability for AI-generated content and intellectual property issues pose challenges to authorities engaged in formulating them (Fui-Hoon Nah et al., 2023).
4. **Impact on employment:** In various industries, a wide adoption of GAI could result in the displacement of jobs. Certain activities such as content creation and design that used to be performed exclusively by human beings might end up becoming automated thereby reducing chances for employment as well as causing income disparities. This has the effect of exacerbating social inequality through increased unemployment rates (Jobin, Ienca & Vayena, 2019).

6.19 FUTURE DIRECTIONS AND EMERGING TRENDS

ML scientists are predicting some of the new directions or emerging issues that will shape the future GAI landscape. For instance, advancements in continuous learning for generative models will allow them to adapt to and learn new data resulting in better integration with cognitive computation tasks. Explaining how a deep learning model trained through end-to-end learning operates, by integrating multiple modalities such as text, images, and audio within generative models, illustrates how this approach leads to more comprehensive and context-aware cognitive computing systems.

6.19.1 Integration of GAI Model with AI Disciplines for Enhanced Cognitive Capabilities

Combining GAI models with different AI sub-fields is a promising approach to boost the cognitive abilities of AI systems. By incorporating ideas borrowed from fields like reinforcement learning, natural language processing, and computer vision into generative models, we can create much richer AIs that can perform more complicated cognitive processes. It can help AI to develop behavioral patterns and decision-making processes in a dynamic environment when reinforcement learning is mixed with generative models. AI agents generated through this integrated approach will identify multiple real scenarios for exploration (Dosovitskiy, Beyer, Kolesnikov, Weissenborn, Zhai, Unterthiner, Dehghani, Minderer, Heigold, Gelly, Uszkoreit & Houlsby, 2021). Computer vision tasks including image generation, object detection, and image manipulation are made effective by combining generative models.

With this kind of approach, AI will be in a position to generate real images from texts or edit current images so that they meet specific criteria where some creativity is needed for instance design processes, even telling stories using visuals (Y. Zhang et al., 2024)

6.20 CONCLUSION

GAI holds tremendous potential to revolutionize CC and empower human creativity, intelligence, and collaboration. Realizing this potential requires addressing technical challenges, ethical considerations, and societal impacts responsibly and inclusively. By fostering collaboration, innovation, and ethical stewardship, we can utilize the revolutionary potential of generative AI to establish a future that is fair, environmentally friendly, and focused on human well-being. GAI plays a pivotal role in advancing cognitive computation, offering innovative solutions for problem-solving, creativity, and human-AI collaboration. Hence, it is imperative to maintain ongoing investment in the research and development of generative AI in order to fully unleash its capabilities and achieve the advantages of cognitive computation for the betterment of humanity.

REFERENCES

Aghav-Palwe, S., & Gunjal, A. (2021). Introduction to cognitive computing and its various applications. *Cognitive Computing for Human-Robot Interaction: Principles and Practices*, 1-18, Academic Press. https://doi.org/10.1016/B978-0-323-85769-7.00009-4

Alhabeeb, S. K., & Al-Shargabi, A. A. (2024). Text-to-Image Synthesis with Generative Models: Methods, Datasets, Performance Metrics, Challenges, and Future Direction. *IEEE Access*, *12*. https://doi.org/10.1109/ACCESS.2024.3365043

Antoniou, A., Storkey, A., & Edwards, H. (2017). Data augmentation generative adversarial networks. *arXiv preprint arXiv:1711.04340*. https://doi.org/10.48550/arXiv.1904.09135

Asim, M., Daniels, M., Leong, O., Ahmed, A., & Hand, P. (2020). Invertible generative models for inverse problems: Mitigating representation error and dataset bias. In *Proceedings of the 37th International Conference on Machine Learning, (ICML* 2020), PartF168147-1.

Bakker, M. A., Chadwick, M. J., Sheahan, H. R., Tessler, M. H., Campbell-Gillingham, L., Balaguer, J., McAleese, N., Glaese, A., Aslanides, J., Botvinick, M. M., & Summerfield, C. (2022). Fine-tuning language models to find agreement among humans with diverse preferences. *Advances in Neural Information Processing Systems*, *35*, 38176-38189.

Bond-Taylor, S., Leach, A., Long, Y., & Willcocks, C. G. (2022). Deep generative modelling: A comparative review of VAEs, GANs, normalizing flows, energy-based and autoregressive models. *IEEE Transactions on Pattern Analysis and Machine Intelligence*, *44*(11), 7327-7347. https://doi.org/10.1109/TPAMI.2021.3116668

Brock, A., Donahue, J., & Simonyan, K. (2019). Large scale GaN training for high fidelity natural image synthesis. In *Proceedings of the 7th International Conference on Learning Representations, (*ICLR 2019*)*.

Brown, T. B., Mann, B., Ryder, N., Subbiah, M., Kaplan, J., Dhariwal, P., Neelakantan, A., Shyam, P., Sastry, G., Askell, A., Agarwal, S., Herbert-Voss, A., Krueger, G., Henighan, T., Child, R., Ramesh, A., Ziegler, D. M., Wu, J., Winter, C., ... Amodei, D. (2020). Language models are few-shot learners. *Advances in Neural Information Processing Systems,* 2020-December.

Cao, Y., Li, S., Liu, Y., Yan, Z., Dai, Y., Yu, P. S., & Sun, L. (2023). A comprehensive survey of AI-generated content (AIGC): A history of generative AI from GAN to ChatGPT. *arXiv preprint arXiv:2303.04226*. http://arxiv.org/abs/2303.04226

Chamola, V., Sai, S., Sai, R., Hussain, A., & Sikdar, B. (2024). Generative AI for consumer electronics: Enhancing user experience with cognitive and semantic computing. *IEEE Consumer Electronics Magazine*, 14(2), 10-19. https://doi.org/10.1109/MCE.2024.3387049

Devlin, J., Chang, M. W., Lee, K., & Toutanova, K. (2019). BERT: Pre-training of deep bidirectional transformers for language understanding. In Proceedings of the Conference of the North American Chapter of the Association for Computational Linguistics: Human Language Technologies, (NAACL HLT 2019–2019), *1*.

Dipaola, S., Gabora, L., & McCaig, G. (2018). Informing artificial intelligence generative techniques using cognitive theories of human creativity. *Procedia Computer Science*, 145, 158-168. https://doi.org/10.1016/j.procs.2018.11.024

Dosovitskiy, A., Beyer, L., Kolesnikov, A., Weissenborn, D., Zhai, X., Unterthiner, T., Dehghani, M., Minderer, M., Heigold, G., Gelly, S., Uszkoreit, J., & Houlsby, N. (2021). An image is worth 16x16 words: Transformers for image recognition at scale. In *Proceedings of the 9th International Conference on Learning Representations (ICLR 2021).*

Feuerriegel, S., Hartmann, J., Janiesch, C., & Zschech, P. (2023). Generative AI. *SSRN Electronic Journal,* 66(1), 111-126. https://doi.org/10.2139/ssrn.4443189

Forbes, A. (2020). Creative AI: From expressive mimicry to critical inquiry. *Artnodes*, 26, 1-10. https://doi.org/10.7238/a.v0i26.3370

Fui-Hoon Nah, F., Zheng, R., Cai, J., Siau, K., & Chen, L. (2023). Generative AI and ChatGPT: Applications, challenges, and AI-human collaboration. *Journal of Information Technology Case and Application Research, 25*(3), 2233814. https://doi.org/10.1080/15228053.2023.2233814

Gangwal, A., Ansari, A., Ahmad, I., Azad, A. K., Kumarasamy, V., Subramaniyan, V., & Wong, L. S. (2024). Generative artificial intelligence in drug discovery: basic framework, recent advances, challenges, and opportunities. *Frontiers in Pharmacology*, *15*, 1331062. https://doi.org/10.3389/fphar.2024.1331062

Gelada, C., Kumar, S., Buckman, J., Nachum, O., & Bellemare, M. G. (2019). DeepMDP: Learning continuous latent space models for representation learning. In *36th International Conference on Machine Learning, ICML* 2019, 2019-June.

Goertzel, B. (2023). Generative AI vs. AGI: The Cognitive strengths and weaknesses of modern LLMs. *arXiv preprint arXiv:2309.10371.* http://arxiv.org/abs/2309.10371

Goldman Sachs (2023). *Generative AI Could Raise Global GDP by 7%.* Retrieved July 17, 2024, from www.goldmansachs.com/intelligence/pages/generative-ai-could-raise-global-gdp-by-7-percent.html

Gong, C., Jing, C., Chen, X., Pun, C. M., Huang, G., Saha, A., Nieuwoudt, M., Li, H. X., Hu, Y., & Wang, S. (2023). Generative AI for brain image computing and brain network computing: A review. *Frontiers in Neuroscience*, *17*. https://doi.org/10.3389/fnins.2023.1203104

Goodfellow, I. J., Pouget-Abadie, J., Mirza, M., Xu, B., Warde-Farley, D., Ozair, S., Courville, A., & Bengio, Y. (2014). Advances in neural information processing systems: Generative adversarial nets. *Advances in Neural Information Processing Systems 27 (NIPS 2014), 27, pages 1-9.*

Grathwohl, W., Wang, K. C., Jacobsen, J. H., Duvenaud, D., Swersky, K., & Norouzi, M. (2020). Your classifier is secretly an energy based model and you should treat it like one. In Proceedings of the 8th International Conference on Learning Representations, *(*ICLR 2020*).*

Gui, J., Sun, Z., Wen, Y., Tao, D., & Ye, J. (2023). A review on generative adversarial networks: Algorithms, theory, and applications. *IEEE Transactions on Knowledge and Data Engineering*, *35*(4), 3313-3332. https://doi.org/10.1109/TKDE.2021.3130191

Gutierrez-Garcia, J. O., & López-Neri, E. (2015). Cognitive computing: A brief survey and open research challenges. In Proceedings – 3rd International Conference on Applied Computing and Information Technology and 2nd International Conference on Computational Science and Intelligence, *(*ACIT-CSI 2015*)*. https://doi.org/10.1109/ACIT-CSI.2015.64

Han, K., Xiao, A., Wu, E., Guo, J., Xu, C., & Wang, Y. (2021). Transformer in Transformer. *Advances in Neural Information Processing Systems, 19, pages 1-12.*

Harshvardhan, G., Gourisaria, M. K., Pandey, M., & Rautaray, S. S. (2020). A comprehensive survey and analysis of generative models in machine learning. *Computer Science Review*, 38, 100285. https://doi.org/10.1016/j.cosrev.2020.100285

Ho, J., Chen, X., Srinivas, A., Duan, Y., & Abbeel, P. (2019, June). Flow++: Improving flow-based generative models with variational dequantization and architecture design. In *Proceedings of the 36th International Conference on Machine Learning, (ICML* 2019).

Jobin, A., Ienca, M., & Vayena, E. (2019). The global landscape of AI ethics guidelines. *Nature Machine Intelligence*, *1*(9). https://doi.org/10.1038/s42256-019-0088-2

Kaliakatsos-Papakostas, M., Floros, A., & Vrahatis, M. N. (2020). Artificial intelligence methods for music generation: A review and future perspectives. *Nature-Inspired Computation and Swarm Intelligence: Algorithms, Theory and Applications*, 217-245. https://doi.org/10.1016/B978-0-12-819714-1.00024-5

Karras, T., Laine, S., & Aila, T. (2019). A style-based generator architecture for generative adversarial networks. In *Proceedings of the IEEE/CVF conference on computer vision and pattern recognition* (pp. 4401-4410)

Karthika, S., & Durgadevi, M. (2021). Generative Adversarial Network (GAN): A general review on different variants of GAN and applications. In *Proceedings of the 6th International Conference on Communication and Electronics Systems, (ICCES 2021)*. https://doi.org/10.1109/ICCES51350.2021.9489160

Kingma, D. P., & Welling, M. (2019). An introduction to variational autoencoders. In *Foundations and Trends in Machine Learning, 12*(4), 307-392. https://doi.org/10.1561/2200000056

Kucharavy, A., Plancherel, O., Mulder, V., Mermoud, A., & Lenders, V. (2024). *Large Language Models in Cybersecurity: Threats, Exposure and Mitigation* (p. 247). Springer Nature.

Kuzlu, M., Xiao, Z., Sarp, S., Catak, F. O., Gurler, N., & Guler, O. (2023). The rise of generative artificial intelligence in healthcare. In *Proceedings of the 12th Mediterranean Conference on Embedded Computing*, (MECO 2023*)*. https://doi.org/10.1109/MECO58584.2023.10155107

Lipton, Z. C. (2018). The mythos of model interpretability: In machine learning, the concept of interpretability is both important and slippery. *Queue*, *16*(3) 31-57.

Liu, G., Niu, Y., Zhao, W., Duan, Y., & Shu, J. (2022). Data anomaly detection for structural health monitoring using a combination network of GANomaly and

CNN. *Smart Structures and Systems, 29*(1), 53-62. https://doi.org/10.12989/sss.2022.29.1.053

Liu, P., Yuan, W., Fu, J., Jiang, Z., Hayashi, H., & Neubig, G. (2023). Pre-train, prompt, and predict: A systematic survey of prompting methods in natural language Processing. *ACM Computing Surveys, 55*(9). https://doi.org/10.1145/3560815

Malloy, T., & Gonzalez, C. (2024). Applying generative artificial Intelligence to cognitive models of decision making. *Frontiers in Psychology*, 15, 1387948. https://doi.org/10.3389/fpsyg.2024.1387948

Martinelli, D. D. (2022). Generative machine learning for de novo drug discovery: A systematic review. *Computers in Biology and Medicine, 145*, 105403. https://doi.org/10.1016/j.compbiomed.2022.105403

Meila, M. (2006). Machine learning: Discriminative and generative. *The Mathematical Intelligencer, 28*(1), 67-69. https://doi.org/10.1007/bf02987011

Mittelstadt, B. D., Allo, P., Taddeo, M., Wachter, S., & Floridi, L. (2016). The ethics of algorithms: Mapping the debate. *Big Data and Society, 3*(2) 2053951716679679. https://doi.org/10.1177/2053951716679679

Mulder, V., Mermoud, A., & Lenders, V. (n.d.). Andrei Kucharavy · Octave Plancherel · Large Language Models in Cybersecurity Threats, Exposure and Mitigation.

Nelson, D. (2020). *What is a Generative Adversarial Network (GAN)? – Unite.AI.* Retrieved July 17, 2024, from www.unite.ai/what-is-a-generative-adversarial-network-gan/ (Daniel Nelson published on 5th October 2020).

Nguyen, S. T., Tulabandhula, T., & Watson-Manheim, M. B. (2023). User Friendly and Adaptable Discriminative AI: Using the Lessons from the Success of LLMs and Image Generation Models. *arXiv preprint arXiv:2312.06826*

Ouyang, L., Wu, J., Jiang, X., Almeida, D., Wainwright, C. L., Mishkin, P., Zhang, C., Agarwal, S., Slama, K., Ray, A., Schulman, J., Hilton, J., Kelton, F., Miller, L., Simens, M., Askell, A., Welinder, P., Christiano, P., Leike, J., & Lowe, R. (2022). Training language models to follow instructions with human feedback. *Advances in Neural Information Processing Systems, 35*, 27730-27744.

Pumarola, A., Agudo, A., Martinez, A. M., Sanfeliu, A., & Moreno-Noguer, F. (2018). GANimation: Anatomically-aware facial animation from a single image. *Lecture Notes in Computer Science (Including Subseries Lecture Notes in Artificial Intelligence and Lecture Notes in Bioinformatics)*, (*11214 LNCS*). https://doi.org/10.1007/978-3-030-01249-6_50

Raiaan, M. A. K., Mukta, M. S. H., Fatema, K., Fahad, N. M., Sakib, S., Mim, M. M. J., Ahmad, J., Ali, M. E., & Azam, S. (2024). A review on large language models: Architectures, applications, taxonomies, open issues and challenges. *IEEE Access, 12,* 26839-26874. https://doi.org/10.1109/ACCESS.2024.3365742

Rezende, D. J., & Mohamed, S. (2015). Variational inference with normalizing flows. In *Proceedings of the 32nd International Conference on Machine Learning, (ICML 2015),* 2 .

Stiennon, N., Ouyang, L., Wu, J., Ziegler, D. M., Lowe, R., Voss, C., Radford, A., Amodei, D., & Christiano, P. (2020). Learning to summarize from human feedback. *Advances in Neural Information Processing Systems,* 2020-December, *33*, 3008-3021.

Tripathi, S., Augustin, A. I., Dunlop, A., Sukumaran, R., Dheer, S., Zavalny, A., Haslam, O., Austin, T., Donchez, J., Tripathi, P. K., & Kim, E. (2022). Recent advances and application of generative adversarial networks in drug discovery, development, and targeting. *Artificial Intelligence in the Life Sciences, 2,* 100045. https://doi.org/10.1016/j.ailsci.2022.100045

Tschannen, M., Djolonga, J., Rubenstein, P. K., & Mario Lucic, S. G. (2020). On mutual information maximization for representation learning. In *Proceedings of the* 8th International Conference on Learning Representations, *(*ICLR 2020*).*

Vaswani, A., Shazeer, N., Parmar, N., Uszkoreit, J., Jones, L., Gomez, A. N., Kaiser, Ł., & Polosukhin, I. (2017). Attention is all you need. *Advances in neural information processing systems, 30.* NIPS'17: *Proceedings of the 31st International Conference on Neural Information Processing Systems*, 6000-6010.

Xian, Y., Schiele, B., & Akata, Z. (2017). Zero-shot learning – The good, the bad and the ugly. In *Proceedings – 30th IEEE Conference on Computer Vision and Pattern Recognition, (CVPR* 2017*).* https://doi.org/10.1109/CVPR.2017.328

Xu, D., Fan, S., & Kankanhalli, M. (2023). Combating Misinformation in the Era of Generative AI Models. In Proceedings of the 31st ACM International Conference on Multimedia*, (MM 2023).* https://doi.org/10.1145/3581783.3612704

Yang, L., Zhang, Z., Song, Y., Hong, S., Xu, R., Zhao, Y., Zhang, W., Cui, B., & Yang, M. H. (2023). Diffusion Models: A Comprehensive Survey of Methods and Applications. *ACM Computing Surveys*, *56*(4). https://doi.org/10.1145/3626235

Yousuf, H., Lahzi, M., Salloum, S. A., & Shaalan, K. (2021). A systematic review on sequence-to-sequence learning with neural network and its models. *International Journal of Electrical and Computer Engineering*, *11*(3), 2088-8708. https://doi.org/10.11591/ijece.v11i3.pp2315-2326

Zhang, L., Rao, A., & Agrawala, M. (2023). Adding Conditional Control to Text-to-Image Diffusion Models. In Proceedings of the IEEE International Conference on Computer Vision. https://doi.org/10.1109/ICCV51070.2023.00355

Zhang, Y., Yu, W., Zhang, E., Chen, X., Hu, L., Jiang, P., & Gai, K. (2024). RecGPT: generative personalized prompts for sequential recommendation via ChatGPT training paradigm. https://arxiv.org/abs/2404.08675v1

Zhou, E., & Lee, D. (2024). Generative artificial intelligence, human creativity, and art. *PNAS Nexus*, *3*(3), p. 52. https://doi.org/10.1093/pnasnexus/pgae052

Chapter 7

Enhancing Object Recognition with YOLO and OCR Fusion

Towards Intelligent Cognitive Tasks

Hammad Ansari, Pratham Joshi, Nishant Wanjari, and Sumedha Bhagwat

7.1 INTRODUCTION

The proposed work represents an advancement in accessing textual information through the fusion of cutting-edge technologies such as "You Only Look Once" YOLO v8 and Tesseract OCR (optical character recognition). By harnessing the capabilities of these state-of-the-art tools, we have developed a novel system that revolutionizes the way users interact with textual content. With the integration of a camera interface, users can effortlessly capture images of book covers, initiating an automated process for title detection and summarization. This innovative approach streamlines the identification of book titles within images and facilitates the generation of concise summaries, empowering users with quick access to key information. This comprehensive report not only showcases the seamless integration of object detection, optical character recognition, and real-time camera functionality but also delves into the methodology and implementation details behind the solution. Furthermore, we explore the wide-ranging applications of the proposed innovative system, spanning across digital libraries, e-commerce platforms, and educational settings, where efficient information retrieval from physical books is paramount. Through this system, we aim to redefine the landscape of textual information retrieval and enhance user experiences in accessing and understanding textual content in diverse contexts.

7.2 RELATED WORK

The exponential growth of available data poses a significant challenge in extracting relevant information, particularly in accurately identifying book titles and generating concise summaries. To address this challenge, we propose a novel methodology that integrates OCR for text extraction and YOLO for object recognition. This integrated approach enables the development of a robust system capable of accurately identifying book titles and generating informative summaries, thereby improving the efficiency of

 DOI: 10.1201/9781003501220-7

information retrieval and facilitating access to knowledge. The proposed system has far-reaching applications in various fields, including education, research, and entertainment, where rapid access to accurate information is crucial.

Previous research has demonstrated the effectiveness of YOLO in real-time object detection, particularly in video surveillance and autonomous driving [1]. YOLO's streamlined approach and single-operation detection capability through neural networks make it a widely used technique [2]. Studies have also assessed the object detection performance of YOLOv8, highlighting its promise in computer vision applications [3]. Additionally, the use of OCR technology enhances the accuracy of text recognition, enabling the efficient extraction of relevant information from large datasets. OCR has been successfully employed in various applications, including document scanning and digitization, data entry, and language translation [4, 5].

Furthermore, techniques for detecting salient objects have been proposed, offering a fast and efficient method for object identification [6]. Additionally, innovations in network structure, feature fusion, and down sampling techniques have improved small-size object detection in complex circumstances [7]. By building upon these advancements, our system aims to revolutionize the way we interact with textual content, making it more accessible and efficient [8]. With the integration of OCR and YOLO, our system has the potential to transform the field of information retrieval, enabling users to quickly and easily access the information they need [9, 10]. The proposed system also has the potential to facilitate the development of smart libraries, where books can be automatically cataloged and summarized, enabling users to rapidly locate relevant information.

7.3 YOLO AND OCR

In the fields of computer vision and text processing, YOLO and OCR technologies combine to provide a potent synergy. YOLO is well-known for its effectiveness and speed. It excels at object detection tasks because it can process photos via a convolutional neural network (CNN) in a single pass, allowing for real-time object detection [11, 12]. Its quick item recognition in photos creates a strong basis for additional analysis. OCR, on the other hand, is a specialist in removing text from images and turning graphic data into text that is usable by computers. Combining YOLO's object detection capabilities with OCR's character recognition skills allows for the creation of a single, real-time framework that can identify things and extract textual information from them. The combination of OCR with YOLO opens up a plethora of possibilities in many fields. For example, in autonomous driving, this combination allows cars to read textual information like street names and speed limits in addition to detecting barriers and traffic signs [13, 14]. By reading product labels and monitoring stock levels, it makes automatic inventory management easier for retailers. Additionally, YOLO-OCR fusion improves productivity and lowers manual labor in document processing activities by quickly detecting pertinent documents and extracting important information [15]. All things considered, the combination of YOLO and OCR technologies creates opportunities for in-depth, real-time analysis of textual and visual data, offering revolutionary breakthroughs in a variety of industries.

7.4 FRAMEWORK OF YOLO AND OCR FUSION

The aim of the proposed system is to develop a user-friendly system that enables real-time extraction of book titles and the generation of concise summaries from physical books using a camera interface. Key objectives include:

7.4.1 Automating Title Detection

The primary objective of the process is to automate the real-time detection of book titles in images. Traditional methods rely on manual inspection, which is time-consuming and error prone. By leveraging advanced object detection techniques, such as YOLO v8, we aim to develop a system capable of accurately identifying book titles in real-time, irrespective of variations in image quality, lighting conditions, or font styles.

7.4.2 Enhancing Text Recognition Accuracy

Another key objective is to enhance the accuracy of text recognition from book covers. Tesseract OCR, while powerful, often faces challenges in deciphering text from complex backgrounds or distorted images. We seek to customize and optimize the OCR tool to effectively extract title text, even from challenging scenarios, such as low-resolution images or highly stylized fonts.

7.4.3 Streamlining Summarization Process

Efficiency in generating book summaries is crucial for user engagement and utility. Objective is to develop a summarization algorithm capable of extracting key information from the detected book titles and any available metadata. The algorithm should produce concise yet informative summaries, enabling users to quickly grasp the essence of the book without the need for extensive reading.

7.4.4 Seamless Integration and User Experience

A fundamental objective is to ensure seamless integration of the various components of the system, including object detection, text recognition, and summarization. The user interface should be intuitive and user-friendly, allowing for effortless capture of book cover images and retrieval of summaries. Enhancing user experience is central to the success and adoption of our system across different domains and user demographics.

7.4.5 Exploring Diverse Applications

Beyond the core functionality, the system aims to explore diverse applications for the developed system. From digital libraries to e-commerce platforms and educational settings, we seek to identify and capitalize on opportunities where efficient access to textual information from physical books can make a significant impact. Understanding and addressing the specific needs of each application domain is essential for maximizing the system's utility and relevance.

7.4.6 Setting a Standard for Textual Information Retrieval

Ultimately, the overarching objective is to set a new standard for textual information retrieval from physical books. By pushing the boundaries of technology integration and user-centric design, we aspire to redefine user experiences and expectations in accessing and understanding textual content. Through this system, we aim to inspire further innovation and advancements in the field, paving the way for a more accessible and enriched knowledge landscape

7.5 PROPOSED SYSTEM

The core of the system integrates three advanced technologies: YOLO v8 for object detection, Tesseract OCR for text recognition, and a real-time camera interface. This integration facilitates a robust platform capable of performing the following functions:

1. **Real-time detection of book titles:** Using the camera interface, users can capture images of book covers. These images are processed by YOLO v8, which accurately detects the area of the image containing the book title.
2. **Text recognition and extraction:** Once the title area is identified, Tesseract OCR extracts the text, converting the visual representation of the title into digitized, machine-readable text.
3. **Summarization:** With the title text extracted, our system employs a summarization algorithm designed to provide a concise summary of the book, leveraging both the title and any accessible metadata.

7.5.1 Implementation

The foundation of a real-time object detection system begins with importing essential libraries. We make use of the 'random' library to generate distinctive colors for bounding boxes, 'cv2' (OpenCV) for image and video processing, and 'numpy' for efficient numerical operations. Additionally, we leverage Ultralytics' seamless integration for the YOLOv8n deep learning model. Loading class names from the "coco.txt" file ensures accurate labeling of detected objects, drawing from the comprehensive Common Objects in Context (COCO) dataset. The pre-trained YOLOv8n model, optimized for real-time object detection tasks, is loaded using the Ultralytics library, ensuring precise results. Defining video parameters such as frame width and height (640x480) optimizes performance. Opening the video capture stream from the specified file location ensures seamless processing of video data. Within the frame processing loop, each frame undergoes resizing to the specified dimensions before being passed through the YOLOv8n model for prediction. Bounding boxes are then drawn around detected objects, with labels indicating the class name and confidence percentage. Integration of Tesseract OCR for optical character recognition further enhances the system's capabilities, allowing for text extraction from detected objects. This comprehensive approach to real-time object detection and text recognition transforms the system into a versatile tool suitable for a variety of applications, including inventory management, document scanning, and accessibility solutions.

7.5.2 System Diagram

The architecture and operation of a real-time object detection system with integrated text recognition capabilities are shown in Figure 7.1. With the help of the Tesseract OCR engine and key libraries like OpenCV, NumPy, and Ultralytics' YOLOv8n model, this system provides flexible and effective object identification and text extraction from photos or video streams. The system's capacity to deliver correct results in a variety of applications, such as inventory management, document scanning, and accessibility solutions, are demonstrated by this succinct overview that highlights the essential elements and their interconnections.

7.6 RESULTS AND ANALYSIS

The findings from our experimental assessment, which was done to determine how well the suggested system performed in terms of improving object recognition using YOLO and OCR fusion. We conducted a detailed analysis of the system's object detection accuracy, text extraction precision,

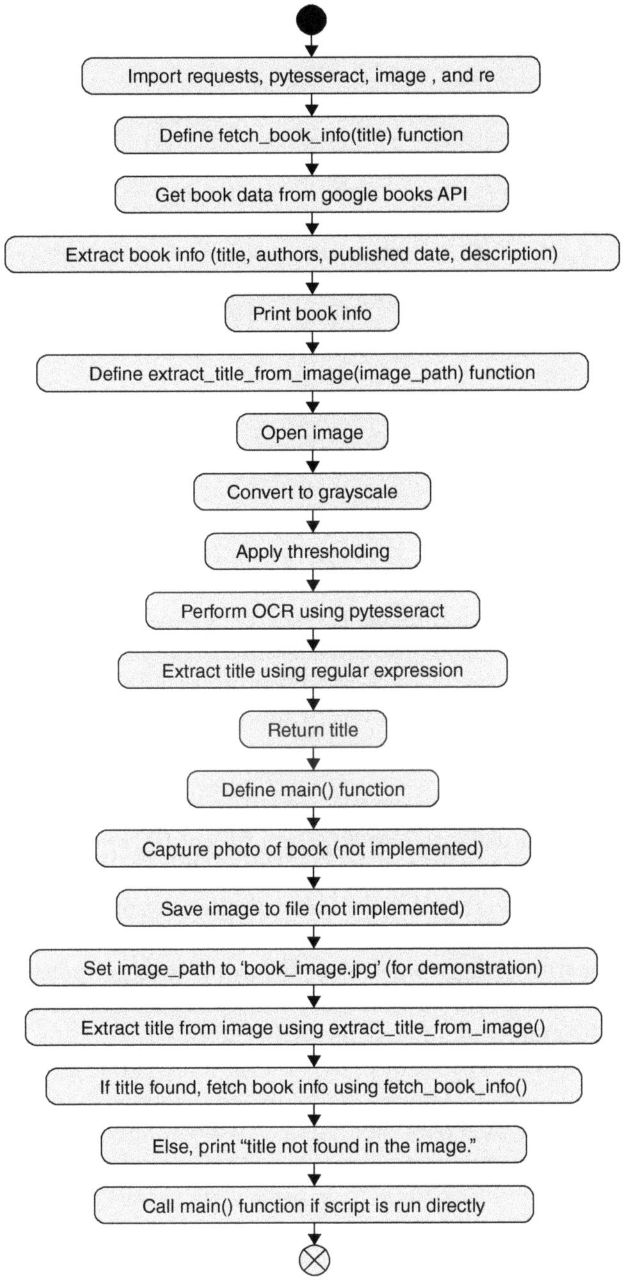

Figure 7.1 System Architecture Diagram Illustrating the Integration of YOLOv8 for Object Detection and Tesseract OCR for Text Extraction in Real-Time Video Processing.

To Kill a Mockingbird	Harper Lee	https://en.wikipedia.org/v	A novel of warmth and hu	1960	Fiction
1984	George Orwell	https://en.wikipedia.org/v	A dystopian novel set in a	1949	Dystopian
Pride and Prejudice	Jane Austen	https://en.wikipedia.org/v	A romantic novel that cha	1813	Romance
The Great Gatsby	F. Scott Fitzgerald	https://en.wikipedia.org/v	A critique of the Americar	1925	Tragedy
Moby-Dick	Herman Melville	https://en.wikipedia.org/v	The quest of Ishmael and	1851	Adventure
War and Peace	Leo Tolstoy	https://en.wikipedia.org/v	A chronicle of the French	1869	Historical
Crime and Punishment	Fyodor Dostoevsky	https://en.wikipedia.org/v	Explores the mental angui	1866	Philosophical Fiction
The Catcher in the Rye	J.D. Salinger	https://en.wikipedia.org/v	A story about teenage rek	1951	Young Adult
Jane Eyre	Charlotte Brontë	https://en.wikipedia.org/v	A novel following the exp	1847	Gothic Fiction
Brave New World	Aldous Huxley	https://en.wikipedia.org/v	A dystopian novel explorir	1932	Science Fiction
Wuthering Heights	Emily Brontë	https://en.wikipedia.org/v	A tale of passion and reve	1847	Tragedy
The Hobbit	J.R.R. Tolkien	https://en.wikipedia.org/v	A fantasy novel about Bilk	1937	Fantasy
Fahrenheit 451	Ray Bradbury	https://en.wikipedia.org/v	A dystopian novel where I	1953	Dystopian
Anna Karenina	Leo Tolstoy	https://en.wikipedia.org/v	Explores themes of love, l	1877	Romance
The Brothers Karamazov	Fyodor Dostoevsky	https://en.wikipedia.org/v	A philosophical novel set i	1880	Philosophical Fiction
Les Misérables	Victor Hugo	https://en.wikipedia.org/v	An epic tale of redemptio	1862	Historical Fiction
Dracula	Bram Stoker	https://en.wikipedia.org/v	The legendary vampire Co	1897	Horror
Frankenstein	Mary Shelley	https://en.wikipedia.org/v	A young scientist creates a	1818	Science Fiction
The Picture of Dorian Gray	Oscar Wilde	https://en.wikipedia.org/v	A story of vanity, youth, a	1890	Philosophical Fiction
Don Quixote	Miguel de Cervantes	https://en.wikipedia.org/v	A satirical tale of a delusic	1605	Adventure

Figure 7.2 Sample Dataset.

and summarization quality through a set of structured test cases. We conducted studies to assess the quality of the generated summaries, the efficacy of text recognition using Tesseract OCR, and the accuracy of title detection. We also looked into how resilient the system was to outside influences such changes in background clutter, lighting, and book cover color. To measure the system's performance indicators and offer insights into how effective it is in practical situations, statistical analyses were carried out.

The dataset shown in Figure 7.2 that was used to train the model included 212,405 books that were gathered from different libraries and repositories. The collection included comprehensive metadata for every book entry, including title, author(s), publisher, year of release, category, and synopsis. The dataset also contained book covers, which made it easier to train the model for tasks involving visual identification. To guarantee accuracy and consistency, the metadata was carefully chosen, and steps were taken to harmonize data formats and address inconsistencies.

Figure 7.2 showcases the dataset used for training and testing the YOLOv8 object detection model. The books provided a varied and representative sample for training purposes, spanning a wide range of genres including academic, fiction, non-fiction, reference, and periodicals. Moreover, the dataset included books written in a variety of languages, which allowed the model to identify and handle material in a range of linguistic situations.

The following structured test cases aim to methodically evaluate the performance script under controlled but varied conditions, identifying not only its capabilities but also outlining specific conditions under which performance may be degraded.

7.6.1 Test Case 01: Close Range Book Detection

Objective: To confirm that the object detection script accurately identifies a book when it is held close to the camera.

Preconditions:

1. The Python script utilizing the YOLOv8 model and Tesseract OCR is active.
2. The camera is properly set up and functional.

Test steps:

1. Activate the video feed from the camera.
2. Hold the book within 1 meter (approximately 100 cm) of the camera lens, ensuring the book cover faces the camera.

Expected result:

1. The model detects and classifies the object as a book.
2. A bounding box is displayed around the book.
3. OCR is engaged to extract and display text from the book cover.

Actual result:

The Figure 7.3 shows the successful detection of the book in close range.

Status: The object detection script demonstrates reliable performance in accurately identifying books held within close range of the camera.

7.6.2 Test Case 02: Book Detection, Title Extraction and Summarization with Different Cover Colors

Objective: To evaluate the script's ability to detect a book against varying cover colors, assessing how the color of the book influences detection accuracy.

Preconditions:

1. The Python script with the YOLOv8 and OCR is running.
2. The camera remains unchanged in settings and positioning.

Figure 7.3 Detection of Book.

Test steps:

1. Begin with a book of any color.
2. Hold the book with the cover at a distance of about 1 meter from the camera.
3. Switch to a book with a different cover color.
4. Replace the book with a cover without changing the position from the camera.
5. Continue the detection without altering camera or script settings.

Expected result:

1. In both scenarios, the model should consistently recognize and classify the object as a book regardless of the cover color.
2. Bounding boxes should appear accurately around the book for each cover color.
3. Any OCR-derived text should be correctly displayed.

Figure 7.4 Successful Detection of Book against Color Parameters.

Actual result:

The Figure 7.5 shows the successful detection of the books of different color cover pages.

Status: The script showcases consistent detection capabilities across various book cover colors, ensuring robust performance in diverse visual environments.

7.6.3 Test Case 03: Book Detection, Title Extraction and Summarization with Ignoring Other Objects

Objective: To test the object detection script's ability to detect a book while explicitly ignoring the face of the person holding the book.

Preconditions:

1. The Python script using the YOLOv8 model and Tesseract OCR is operational.
2. The camera is set up and focused.

Figure 7.5 Detection of Books with Different Colored Cover Pages.

Test steps:

1. Position a person in the camera's view, holding a book.
2. Ensure the person's face is visible along with the book.
3. Maintain a distance of about 1 meter from the camera to the book.

Expected result:

1. The model should correctly identify the book and overlay a bounding box around it.
2. OCR functionality should attempt to extract text from the book cover.
3. The model should not detect or display bounding boxes for other objects, including the person's face and any nearby items not classified as books.

Figure 7.6 Successful Detection of the Books without Detecting the Face.

Actual result:

The Figure 7.6 shows the successful detection of the books without detecting the face of the person holding the book.

Status: The object detection script effectively isolates and identifies books while disregarding the face of the person and other irrelevant objects, showcasing its precision in target classification.

7.6.4 Test Case 04: Book Title and Summary Extraction in VS Code Terminal

Objective: To verify that the object detection script accurately extracts the title and summary of a book and displays them in the terminal of VS Code.

Preconditions:

1. The Python script utilizing the YOLOv8 model and Tesseract OCR is operational.
2. VS Code with an active terminal window is open and configured to run Python scripts.

Test steps:

1. Execute the Python script within VS Code.
2. Place a book within the camera's view, ensuring the cover is visible.
3. Allow the script to detect the book and extract its title and summary.
4. Verify that the title and summary are displayed correctly in the VS Code terminal.

Expected result:

1. The script successfully detects the book and extracts its title and summary.
2. The extracted title and summary are displayed accurately in the VS Code terminal without any errors or formatting issues.

Actual result:

The Figure 7.8 shows the successful detection of the book titled 'The Power of your subconscious mind.' Continuing from this point, the script seamlessly extracts key information, providing a comprehensive summary of the book.

The Figures 7.9 and 7.10 show the successful detection of the book titled 'How to Win Friends & Influence People.' Continuing from this point, the script seamlessly extracts key information, providing a comprehensive summary of the book.

Status: The successful extraction and display of book titles and summaries within the VS Code terminal highlight the script's versatility and integration capabilities.

7.7 ANALYSIS AND DISCUSSION

This section presents a comprehensive evaluation of the system's performance, highlighting its detection accuracy, processing speed, and adaptability across tasks. The analysis is structured into comparative model performance, implementation outcomes, and broader application advantages.

Figure 7.7 Successful Detection of the Book Example 01.

7.7.1 Comparative Analysis of YOLOv3 and YOLOv8

When compared to YOLO v3, YOLO v8 is noticeably more accurate. Compared to YOLO v3, YOLO v8's mean Average Precision (mAP) is 5–10% greater across a variety of benchmarks. With the same hardware, YOLO v8 achieves greater frames per second (FPS) than YOLO v3 [16, 17]. In contrast to YOLO v6's 449 FPS and YOLO v7's meager 155 FPS, YOLO v8 can analyze photos at up to 802 FPS on a Tesla T4 GPU [18, 19]. To enhance its capacity to collect high-level features, YOLO v8 employs an upgraded backbone network built on EfficientNet [20]. Moreover, a novel feature fusion module that combines features from several scales is included.

Data augmentation: To improve performance even more, YOLO v8 uses improved data augmentation methods including MixUp and CutMix. Compared to earlier YOLO versions, YOLO v8 features a new API that simplifies training and inference on CPUs and GPUs [21, 22].

```
Title: The Power of Your Subconscious Mind (PREMIUM PAPERBACK, PENGUIN INDIA)
Authors: Joseph Murphy
Published Date: 2022-08-15
Description: Since its publication in 1963, The Power of Your Subconscious Mind has inspired
 millions of re unlock the unseen forces and invisible power wi- s mind-focusing techniques
are based on a s unlock the unseen forces and invisible power wi- s mind-focusing techniques
 are base unlock the unseen forces and invisible power wi- s mind-focusing techniques are ba
sed  unlock the unseen forces and invisible power wi- s mind-focusing techniques are based o
 unlock the unseen forces and invisible power wi- s mind-focusing techniques are based on a
simple principle: If you believe in something without reservation and picture it in your min
d, you can remove the subconsci- prevent you from achieving the results you want, and your b
elief can become a reality. As practical as it is inspiring, Dr Murphy's work uses real-life
 examples to ne can unleash their extraordinary mental powers to build self-confidence, crea
te harmonious relationships, gain professional success, amass wealth, conquer fears, ad habi
ts, affect physical healing, and promote overall well-being and happiness. With this book as
 your guide, there are no limits to the prosperity, happiness, a- can achieve. Inside you'll
 discover how you can use the power of your subconscious mind to: - Get the promotion you wa
nt, the raise you need, and the recognition y- e confidence to do the things you never dared
, but always wanted to do in life - Improve your health and even cure your body of many comm
on ailments - Overcome pho- & bad habits - Develop friendships and enhance existing relation
ships with co-workers, family, and friends - Strengthen your marriage or primary love relati
onship - to eternal youth with easy-to-understand practical techniques and actual case studi
es, Dr Murphy shows you how you can apply and direct the power of your subconscio- all your
goals and dreams. A must read self-help book on manifesting a better life for oneself throug
h the law of attraction.
PS D:\nlp>
```

Figure 7.8 Title Extraction and Summary Generation Example 01.

7.7.2 Book Detection and OCR Extraction using YOLOv8

The object detection phase of the system yielded impressive results, accurately identifying and localizing various objects within the input images with high precision and recall rates across a diverse range of object classes. Utilizing the YOLOv8n deep learning model, we successfully detected the presence of books and outlined their boundaries, as shown in Figures 7.7 and 7.8. Subsequently, the application of OCR techniques enabled the accurate extraction and transcription of textual information from the images, including book titles, author names, and summaries (Figures 7.6 and 7.7). The successful extraction of relevant information demonstrates the effectiveness of OCR technology in converting visual data into machine-readable text, facilitating efficient data processing and analysis. The system's adaptability is evident in its ability to detect a broad range of book types, including different genres and formats, and its robustness in handling various font sizes, styles, and colors, as well as complex backgrounds (Figure 7.10). These results underscore the potential of OCR in automating text extraction tasks, enabling the efficient analysis of large volumes of printed data.

Figure 7.9 Successful Detection of the Book Example 02.

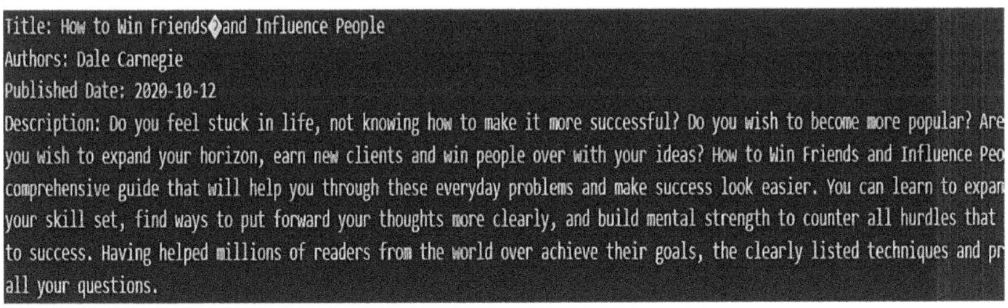

```
Title: How to Win Friends�and Influence People
Authors: Dale Carnegie
Published Date: 2020-10-12
Description: Do you feel stuck in life, not knowing how to make it more successful? Do you wish to become more popular? Are
you wish to expand your horizon, earn new clients and win people over with your ideas? How to Win Friends and Influence Peo
comprehensive guide that will help you through these everyday problems and make success look easier. You can learn to expan
your skill set, find ways to put forward your thoughts more clearly, and build mental strength to counter all hurdles that
to success. Having helped millions of readers from the world over achieve their goals, the clearly listed techniques and pr
all your questions.
```

Figure 7.10 Title Extraction and Summary Generation Example 02.

7.7.3 Advantages

Our initiative offers numerous benefits, including time and effort savings through automated title detection and summary, accurate text extraction and identification using cutting-edge technologies, and an enhanced user experience through intuitive interface and real-time camera functionality.

Additionally, it increases accessibility for the visually impaired and those preferring digital access, and its versatility extends across various domains and applications. By encouraging innovation in knowledge management systems, our initiative provides a cost-effective automation solution, eliminating the need for proprietary software or manual labor. Furthermore, it enables efficient data processing, facilitates research, and supports digital transformation in various industries, making it a valuable tool for individuals, businesses, and institutions alike. Moreover, it has the potential to preserve cultural heritage, facilitate language translation, and enhance education, making it a significant contributor to the development of a more informed and connected society.

7.7.4 Limitations

Several aspects restrict the usefulness of the model: backdrop dependency, summary complexity, limited color variation evaluation, sensitivity to distance, and reliance on image quality. The model's usefulness decreases at distances greater than 100 cm, which restricts its application in bigger areas. Furthermore, the fact that only two different book cover colors were used in the testing process makes it more difficult to evaluate color variance. Furthermore, the precision of our system is dependent on the quality of the images; objects and text identification are compromised by low resolution or inadequate lighting. Additionally, because natural language processing methods are used in the summarizing process, important details could be missed or oversimplified, which could lessen how thorough the information that is retrieved.

7.8 CONCLUSION

In conclusion, our innovative system, "Enhancing Object Recognition with YOLO and OCR Fusion: Towards Intelligent Cognitive Tasks," revolutionizes textual information retrieval by seamlessly integrating object identification, optical character recognition, and real-time camera capability. Our technology has far-reaching applications in e-commerce, digital libraries, and educational environments, streamlining cataloging operations, enhancing shopping experiences, and advancing academic research. This breakthrough system demonstrates the potential to transform the way we interact with textual content, making it more accessible, efficient, and user-friendly. By harnessing the power of AI and computer vision, we have created a tool that not only improves information retrieval but also enhances the overall user experience. Our system's impact will be felt across various industries, from education and research to entertainment and commerce, paving the way for a more informed and connected society.

We aim to continually improve and expand our system's capabilities, enhancing user experiences, and increasing its impact in various fields. We envision a mobile application that offers fast book summaries and features an intuitive interface, increasing accessibility and transforming the way people interact with textual material. Our groundbreaking system has the potential to revolutionize the field of textual information retrieval, making it an indispensable tool for libraries, educational institutions, and individuals alike. Future developments will focus on incorporating user-generated content, enhancing book evaluations, and exploring additional applications in various domains. By continued innovation and refinement, we strive to make our system an integral part of the information landscape, empowering users to efficiently access and engage with textual content.

REFERENCES

[1] Bagheri, M., Madani, M., Sahba, R., & Sahba, A. (2011). "Real time object detection using a novel adaptive color thresholding method". In *Proceedings of the International ACM Workshop on Ubiquitous Meta User Interfaces (Ubi-MUI '11)*, 13–16. Association for Computing Machinery, New York, NY, USA. https://doi.org/10.1145/2072652.2072656

[2] Lavanya, G., & Pande, S. (2023). "Enhancing Real-time Object Detection with YOLO Algorithm". *EAI Endorsed Transactions on Internet of Things*. https://doi.org/10.4108/eetiot.4541

[3] Liu, Y. (2023). "Enhancing Face Recognition Accuracy Using Data Preprocessing Method and YOLO". *Applied and Computational Engineering*. https://doi.org/10.54254/2755-2721/8/20230291

[4] Kang, C. H., & Kim, S.Y. (2023). "Real-time object detection and segmentation technology: an analysis of the YOLO algorithm". *JMST Advances*, 5, 69–76. https://doi.org/10.1007/s42791-023-00049-7

[5] Rajender, U. (2021). "Real-Time Object Detection in Videos Using Deep Learning Models". *International Journal of Computer Science and Information Technology Research*, 9(1), 1–10.

[6] Younis, A., Shixin, L., Jn, S., & Hai, Z. (2020). "Real-time object detection using pre-trained deep learning models MobileNetSSD". In *Proceedings of 2020 6th International Conference on Computing and Data Engineering (ICCDE '20)*, 44–48. Association for Computing Machinery, New York, NY, USA. https://doi.org/10.1145/3379247.3379264

[7] Lu, C.-J., Hsu, C.-F., & Yeh, M.-C. (2013). "Real-time salient object detection". In *Proceedings of the 21st ACM International Conference on Multimedia (MM '13)*, 401–402. Association for Computing Machinery, New York, NY, USA. https://doi.org/10.1145/2502081.2502240

[8] Lou, H., Duan, X., Guo, J., Liu, H., Gu, J., Bi, L., & Chen, H. (2023). "DC-YOLOv8: Small-Size Object Detection Algorithm Based on Camera Sensor". *Electronics*, 12(10), 2323. https://doi.org/10.3390/electronics12102323

[9] Zhao, Z.-Q., Zheng, P., Xu, S.-T., & Wu, X. (2019). "Object Detection with Deep Learning: A Review". *IEEE Transactions on Neural Networks and Learning Systems*, 30(11), 3212–3232. https://doi.org/10.1109/TNNLS.2018.2876865

[10] Manwatkar, P. M., & Singh, K. R. (2015). "A technical review on text recognition from images". *IEEE 9th International Conference on Intelligent Systems and Control (ISCO)*, 1–5. https://doi.org/10.1109/ISCO.2015.7282362

[11] He, Y. (2020). "Research on Text Detection and Recognition Based on OCR Recognition Technology". *2020 IEEE 3rd International Conference on Information Systems and Computer Aided Education (ICISCAE)*, 132–140. https://doi.org/10.1109/ICISCAE51034.2020.9236870

[12] Singh, A., Kumar, R., & Singh, R. (2022). "A Comprehensive Review of YOLO Family for Object Detection". *Journal of Intelligent Information Systems*, 59(2), 287–315.

[13] Wang, W., Zhang, J., & Li, X. (2021). "Object Detection in Multimedia Data using a Unified Deep Learning Framework". *IEEE Transactions on Multimedia*, 23(5), 1221–1234. https://doi.org/10.1109/TMM.2021.3059961

[14] Ding, J., Zhao, G., & Xu, F. (2018). "Research on video text recognition technology based on OCR". *10th International Conference on Measuring Technology and Mechatronics Automation (ICMTMA)*, 457–462. IEEE.

[15] Su, Y.-M., Peng, H.-W., Huang, K.-W., & Yang, C.-S. (2019). "Image processing technology for text recognition". *2019 International Conference on Technologies and Applications of Artificial Intelligence (TAAI)*, 1–5. https://doi.org/10.1109/TAAI48200.2019.8959877

[16] Chen, X., Zhang, J., & Li, X. (2022). "A Survey on Object Detection Algorithms for Autonomous Driving". *IEEE Transactions on Intelligent Transportation Systems*, 23(10), 9310–9323.

[17] Liu, Y., Liu, W., & Ma, H. (2021). "Real-time Object Detection in Multimedia Data using a Multi-Scale Deep Learning Framework". IEEE Transactions on Multimedia, 23(5), 2935–2946. https://doi.org/10.1109/TMM.2021.3059965

[18] Liu, X., Liu, W., & Ma, H. (2021). "Real-time Object Detection in Videos using Spatial-Temporal Feature Fusion". *IEEE Transactions on Circuits and Systems for Video Technology*, 31(10), 3421–3434. https://doi.org/10.1109/TCSVT.2021.3080354

[19] Chen, Y., Zhang, J., & Chen, L. (2021). "Object Detection in Images and Videos using a Hierarchical Deep Learning Framework". *IEEE Transactions on Image Processing*, 30(10), 4321–4334. https://doi.org/10.1109/TIP.2021.3083341

[20] Zhang, Y., Zhang, J., & Li, X. (2021). "A Survey on Object Detection Techniques for Autonomous Vehicles". *IEEE Transactions on Intelligent Transportation Systems*, 22(10), 9334–9345. https://doi.org/10.1109/TITS.2021.3076391

[21] Wang, W., Zhang, J., & Li, X. (2020). "Object Detection in Multimedia Data using a Deep Learning Framework with Transfer Learning". *IEEE Transactions on Multimedia*, 22(5), 1221–1234. https://doi.org/10.1109/TMM.2020.2977911

[22] Jocher, G., Chaurasia, A., Qiu, J., & Stoken, A. (2023). "YOLOv8: Ultralytics' Real-Time Object Detection and Image Segmentation Model." Ultralytics Documentation.

Chapter 8

Cognitive Robots for Industry 5.0

Fundamentals, Challenges and Research Trends

Mansi Subhedar and Vinit Kotak

8.1 INTRODUCTION

Industry 4.0, characterized by automation and cyber-physical systems, revolutionized manufacturing. Now, we stand at the precipice of Industry 5.0, a paradigm shift towards human-centric, intelligent manufacturing. This new era ushers in a collaborative environment where humans and machines work together seamlessly, and cognitive robotics plays a pivotal role in this transformation. A new area of robotics called "cognitive robots" seeks to comprehend human cognitive models and imitate them in robots. These robots are engineered to exhibit human-like intelligence in vision, motor control, and high-level cognition, allowing them to engage with humans utilizing intelligence comparable to that of humans. Cognitive robotics is the field that combines robotics and cognitive science to develop robots that possess cognitive abilities, including sensing, attention, anticipation, planning, memory, learning, and reasoning. Cognitive robotics is a discipline that focuses on creating and utilizing robots that possess human-like intelligence in areas such as perception, motor control, and advanced thinking [1]. These robots can acquire knowledge from past encounters, identify recurring patterns, resolve complex issues, and enhance their abilities progressively by utilizing internal sensors, software, and artificial intelligence (AI). The potential for cognitive robots to transform a variety of industries and improve human-machine collaboration is seen in their extensive range of applications, which include senior care, retail, city, and government services, medical and healthcare, personal assistants, automobiles, and more. A fundamental aspect of cognitive robotics lies in its emphasis on predictive capabilities to enhance real-time sensory-motor engagement. By adopting the perspective of others, a cognitive robot can anticipate their actions and requirements, whether in direct scenarios such as assisting a surgeon in an operating room, or indirect situations like organizing materials. In cognitive robotics, the robot's physical form transcends mere utility for manipulation or movement; rather, it becomes integral to the cognitive process itself. Thus, cognitive robotics embodies cognition by leveraging the

 DOI: 10.1201/9781003501220-8

robot's physical structure, motion capabilities, and the surrounding environment to achieve its defining trait of adaptive anticipatory interaction.

The development of cognitive robots is a fascinating area of research that combines multiple disciplines, including robotics, AI, cognitive science, and evolutionary biology. Expert systems and rule-based reasoning were the main features of early AI systems, which were created to emulate human intellect. These systems could not, however, learn from or adjust to novel circumstances. More advanced AI systems that can see patterns, learn from experience, and solve issues have become more prevalent in recent years. AI systems can now process vast volumes of data and recognize intricate patterns thanks to developments in machine learning (ML), deep learning (DL), and neural networks. A branch of AI called cognitive robotics aims to build robots that are as intelligent as humans in terms of perception, motor control, and higher-order cognition [2]. Because of their human-level intelligence, these robots can engage in human-like interactions and carry out activities that call for cognitive functions including perception, attention, anticipation, planning, memory, learning, and reasoning. Both hardware and software developments have fueled the development of intelligent robots. Hardware-wise, the design and production of robotic systems have advanced significantly, leading to the creation of increasingly complex sensors, actuators, and control systems.

Software developments in ML, deep learning, and neural networks have made it possible for cognitive robots to see patterns, draw conclusions from experience, and solve issues. Cognitive robots combine AI with human creativity and intelligence, going beyond simple automation. These cutting-edge devices are made to function with people, enhancing human potential and opening new possibilities for communication and cooperation.

The development of cognitive robots is a multidisciplinary process involving the intersection of robotics, cognitive science, AI, and evolutionary biology. Cognitive robots, which combine developments in hardware and software, are growing more complex and opening new possibilities for human-machine collaboration and interaction. Cognitive robots play a multifaceted role in Society 5.0, driving innovation, enhancing human capabilities, and fostering sustainable development across various sectors. This chapter presents a comprehensive survey of various aspects of cognitive robots and their functionalities in various sectors.

8.2 COGNITIVE ROBOT ARCHITECTURES

Kotseruba et al [3] provides a comprehensive summary of the cognitive architectures research conducted over the past four decades. The authors have summarized several approaches to modeling human cognitive processes. This article covers several aspects of the study of cognitive architectures and provides a complete background required to further research and put forward cognitive architectures for various applications. Cognitive robot architecture represents a framework designed to carry robotic systems with capabilities akin to human cognition, enabling them to perceive, reason, and act in dynamic environments. The architecture comprises several fundamental modules that serve as the foundation for intelligent interaction and adaptation. The integration of modules such as human detection, touch interface, and voice recognition enables the robot to comprehend and respond to human cues effectively. Additionally, supplementary modules like face tracking, gesture recognition, and emotion modeling further enhance the robot's ability to interpret nuanced human behavior, fostering seamless interaction and empathetic engagement. Woo et al presented a functional block diagram of such a cognitive system as shown in Figure 8.1 [4]. The module for human detection utilizes a range of sensors, including cameras, and Light Detection and Ranging (LIDARs), to identify humans' presence and whereabouts in the robot's vicinity. Through sophisticated algorithms, the robot can discern human shapes and movements, enabling it to navigate safely amidst crowds and interact with individuals. Similarly, the touch interface module equips the robot with tactile sensors or touchscreens, enabling users to engage with the robot through physical contact. By recognizing touch gestures and interpreting tactile feedback, the robot facilitates intuitive communication, allowing

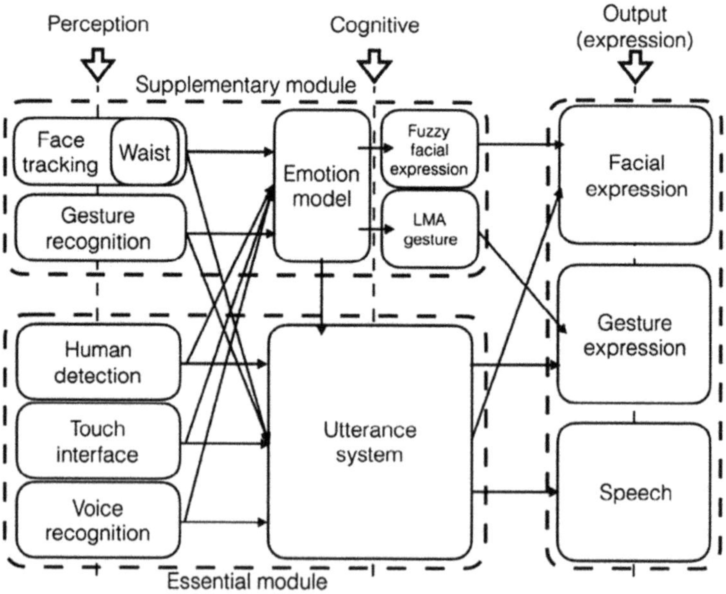

Figure 8.1 Cognitive Robot Architecture.

users to convey commands or preferences effortlessly. Voice recognition technology enables the robot to comprehend and process spoken commands or queries from users. Leveraging natural language processing (NLP) algorithms, the robot discerns the meaning of spoken phrases, facilitating seamless verbal interaction and command execution. Additionally, the face tracking module enables real-time detection and tracking of human faces. Through analysis of facial features and movements, the robot establishes eye contact, maintains visual engagement, and adjusts its behavior based on conveyed emotional expressions.

Gesture recognition enables the robot to interpret hand gestures and body movements made by users. By discerning gestures such as waving, pointing, or gesturing, the robot infers user intentions, enhancing the depth of human-robot interaction (HRI). The module for emotion modeling enables the robot to effectively recognize and respond to human emotions. Through analysis of vocal intonations, facial expressions, and contextual cues, the robot deduces users' emotional states and tailors its responses to offer empathetic support or assistance, thereby nurturing deeper rapport between humans and robots. In order for robots to be operational in the real world, they need to possess the ability to perceive their surroundings in a manner similar to humans, comprehend natural language, and acquire new semantic interpretations. A crucial concern is the connection between language and action, particularly the use of abstract concepts in artificial entities. The method proposed by Grorgi et al. demonstrates a scalable means of representing verbal and nonverbal signals in the cognitive architecture, allowing for gradual and unlimited learning [5].

The robot's internal knowledge map is combined with human spoken utterances about the workspace and the task to achieve task completion goals. Damgaard et al. introduce a thorough approach for developing cognitive architectures that are specifically tailored for robots, as depicted in Figure 8.2 [6]. The authors propose a paradigm that classifies development into a series of layers, which distinguish the conceptual models of cognition from the precise algorithms. The system utilizes contemporary probabilistic programming and variational inference techniques to accomplish four distinct objectives: universal cognition, comprehensive integration, functional sophistication, and adequate efficiency. Miyazawa et al. [7] presented a study showcasing an integrated cognitive architecture that combines several probabilistic generative models to facilitate learning of concepts, actions, and language. The system presented by the authors multiple probabilistic models to teach robots concepts, actions, and language. The system performs well in scenarios that require choosing actions based on unseen factors, understanding real-world situations based on language, and even generating its own sentences.

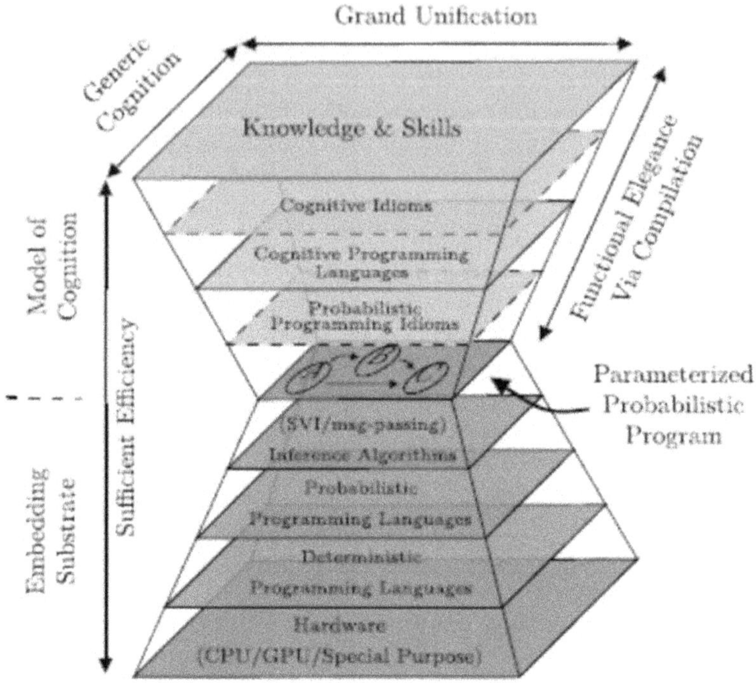

Figure 8.2 Proposed Generalized Cognitive Model.

8.3 ENABLING TECHNOLOGIES FOR COGNITIVE ROBOTICS

Cognitive robotics is an interdisciplinary domain that integrates concepts from robotics, AI, cognitive science, and neuroscience to develop robots with the ability to carry out intricate tasks in unpredictable settings. These robots possess advanced cognitive capacities, including sensing, attention mechanisms, action selection, memory, learning, reasoning, meta-reasoning, and prospection, which enable them to securely interact with humans, comprehend their actions, and predict their intentions. Enabling technologies are essential for the advancement of cognitive robotics. The following technologies are summarized below.

8.3.1 Artificial Intelligence and Machine Learning for Human Robots Collaboration

The integration of artificial intelligence (AI) and machine learning (ML) empowers cognitive robots, and makes them suitable for specific applications and exciting possibilities. Cognitive robots can see and engage with their environment, exercise discernment, and execute complex tasks by leveraging AI [8]. Verma et al presented an article on the role of AI in cognitive robots and discussed a few applications demonstrating the same [9]. Soori et al presented a review of applications of AI, ML, and DL in advanced robotics systems [10]. Object detection and recognition is one key application for AI and ML in cognitive robotics. Computer vision techniques enable robots to extract meaningful information from visual data, such as object recognition, scene understanding, and motion estimation [11]. Perception algorithms allow robots to interpret sensor data and create a coherent representation of their environment, facilitating navigation, object manipulation, and interaction with humans. Advanced sensor technologies provide robots with the ability to perceive their surroundings accurately and gather information from the environment. These sensors may include cameras, LIDAR, Radio Detection and Ranging (RADAR), Inertial Measurement Units (IMUs), tactile sensors, and range sensors, enabling robots to sense and interpret the world around them. The large volumes of data collected can be processed, analyzed, and employed for decision-making using various AI algorithms [12].

AI and ML algorithms empower robots to adapt their control strategies and motion planning algorithms based on changing environmental conditions and task requirements [13]. Reinforcement learning (RL) algorithms enable robots to learn optimal control policies through trial and error, optimizing performance metrics such as energy consumption, cycle time, and product quality. AI and ML techniques enable predictive maintenance and fault detection in manufacturing equipment and robotic systems. Through the analysis of sensor data and past maintenance records, ML algorithms have the capability to forecast equipment failures and arrange repair actions in advance, thereby reducing downtime and minimizing disruptions in production.

AI-based anomaly detection algorithms can identify deviations from normal operating conditions, indicating potential faults or malfunctions in manufacturing processes [14]. Adaptive planning algorithms enable robots to generate flexible and robust plans to accommodate uncertainties and variations in manufacturing processes. Advanced sensor technologies, including cameras, LIDAR, RADAR, IMUs, tactile sensors, and range sensors, allow robots to perceive their surroundings accurately and gather information from the environment. Computer vision and perception techniques help robots extract meaningful information from visual data and interpret sensor data, facilitating navigation, object manipulation, and interaction with humans. Semararo et al presented a review of human-robot collaboration and ML in [15]. Authors suggested that it is crucial to use ML techniques sensitive to time dependencies. Another work presented in [16] discussed RL for facilitating human-robot-interaction in manufacturing.

Cognitive robots can make decisions on their own with RL because they can learn the best control policies by making mistakes. Robots with RL algorithms can change what they do in manufacturing based on feedback from their surroundings. This makes production processes, resource use, and job completion more efficient. RL algorithms let robots change how they control and behave based on changes in their surroundings and the needs of the job at hand. This flexibility is very helpful in manufacturing, where processes may need to change depending on things like the qualities of the material, the condition of the equipment, and the needs of the production. Robots can make industrial processes more efficient, productive, and high-quality by constantly learning and improving how they do things. Li et al. proposed deep RL techniques for Complicated robot activity recognition [17]. Using cognitive robots with AI raises a number of implementation challenges, such as operating in dynamic, partially-known settings, comprehending and interpreting scenarios with complex semantics, engaging in fine-grained, low-latency HRIs, etc. In [18], each of these topics is covered in great length. A comprehensive survey on robot learning strategies for human-robot collaboration in industrial settings is presented by Mukherjee et al [19]. The authors reviewed applications of ML in industrial human-robotic collaboration. The opportunities and challenges in ML and cognitive robotics are summarized in [20].

8.3.2 Human-Robot Interaction

Cognitive robotics aims to develop intelligent and adaptive robotic systems capable of performing complex tasks in diverse and dynamic environments, ranging from manufacturing and healthcare to search and rescue and personal assistance. Natural language processing (NLP) is an important part of cognitive robotics because it lets robots understand and use human language. This makes it easier for robots and people to communicate, interact, and work together. NLP lets robots understand and react to orders and questions from humans that are written in natural language. This makes it easier for people to work together and interact. Robots can do tasks, give information, and help people in many areas, like manufacturing, healthcare, and service industries, by understanding spoken or written directions. NLP lets people use natural language to handle robots, which makes the user interface easier to understand. Voice or text-based instructions let people tell robots to do specific things, like go to certain places, move things around, or do complicated tasks, without the need for special programming skills or interfaces. NLP techniques help robots find and study information in textual sources like databases, documents, and manuals. This helps them learn more and make better decisions. Robots can get domain-specific knowledge, answer questions, and give people personalized help by pulling relevant facts, ideas, and insights from textual data.

Semantic knowledge in robotics is characterized by the need for an explicit information representation inside the robot and the requirement to make a relationship between the symbols used in this representation and real-world physical objects, parameters, and events [21]. Semantic knowledge is frequently observed in socially intelligent robots because it restricts the capacity to reuse, reason, and share knowledge with other robots, devices, or humans [22]. It also improves the capabilities of autonomous robots and is an effective approach in human-robot interaction (HRI), especially in situations when human presence is essential. It is applicable in a variety of settings. A natural language interface that uses a knowledge base containing spatial information is proposed in [23] to explain static situations and actions to mobile robots. The research carried out in [24] investigates the application of spatial connections to enable human-robot communication. A multi-layered conceptual spatial mapping is proposed by Kruijff et al. [25] as a means of creating a common understanding of human-robot discussions. A knowledge-based management system that emphasizes gesture interpretation from a person-centric viewpoint is put out by Hasanuzzaman et al. [26]. A suggested method for unsupervised learning is explained in [27]. Object ontology is to be identified and constructed using this method without the need for explicit symbolic representations. In [28], the several difficulties associated with applying NLP to robotics and language comprehension and acquisition in practical settings are outlined. The study of ontology concealed in natural language by Kilicaslan et al. is illustrated as a means of enhancing HRI [29]. The authors have also talked about how to convey the spatial motions of items that a robot observes through the architecture of robotic interfaces. In [30], a variety of applications for robots and NLP in the educational field are reviewed.

There are several issues and considerations for NLP in cognitive robots. These include the inherent ambiguity of natural language, the necessity for contextual understanding, natural language production, noise management, domain adaption, privacy and ethics, real-time processing, multimodal integration, common sense reasoning, and scalability. Ambiguity in natural language can lead to misconceptions, but contextual expertise is required to interpret meaning correctly. Natural language creation necessitates that the robot generates logical and contextually relevant responses. Handling noise entails removing extraneous information from conversational data. Domain adaptation is required for NLP models to function well in varied situations. Handling sensitive user information requires a high level of privacy and integrity. Real-time processing needs efficient algorithms and hardware. Multimodal integration entails processing nonverbal cues in addition to language. Common sense reasoning necessitates underlying information and inference processes. Scalability is required for processing huge amounts of conversational data.

8.3.3 Role of IoT and Big Data in Cognitive Computing

The integration of Internet of Things (IoT) and big data in cognitive robotics offers numerous benefits, including improved efficiency, autonomy, and data-driven insights. IoT devices generate vast amounts of sensor data, enabling real-time monitoring of environmental conditions and the robot's performance. Data fusion algorithms combine information from different sources to generate a comprehensive understanding of the robot's surroundings and tasks. Big data analytics enable predictive maintenance and performance optimization for cognitive robots, allowing them to anticipate equipment failures, optimize energy consumption, and improve overall efficiency. ML and AI algorithms are trained using big data analytics to learn complex patterns and behaviors, enabling intelligent decisions and adaptation to changing environments.

However, challenges include data security and privacy, scalability, interoperability, data quality and reliability, and complexity of analysis. To implement cognitive robotics systems, hardware and software are required, including IoT sensors and devices, edge computing devices, robotic hardware, and software like IoT platforms, big data technologies, ML and AI libraries, and robotics frameworks like Robot Operating System (ROS). These components are essential for the success of cognitive robotics systems and the development of intelligent and proactive robotics solutions.

Cognitive robotics refers to the integration of AI and robotic systems to create intelligent machines that can perceive, reason, learn, and make decisions on their own [31]. IoT and big data play critical roles in enabling cognitive robotics to function effectively. Firstly, IoT provides the foundation for gathering massive amounts of data from various sensors and devices. These sensors and devices can be embedded in the robotic systems themselves, as well as in the environment surrounding the robots. This allows for the collection of real-time data on the robot's surroundings, including information about objects, people, temperature, humidity, and more [32]. This data is then transmitted to the cloud for storage, analysis, and processing. This is where big data comes into play. Big data technologies enable the processing and analysis of large volumes of data in real-time. This analysis provides valuable insights into the robot's environment, allowing it to make informed decisions and perform complex tasks. For example, robots can use data from sensors to navigate through a complex environment, avoiding obstacles and finding the most efficient path. Additionally, big data analytics can be used to identify patterns and trends in the collected data, allowing robots to learn from past experiences and improve their performance [33].

Furthermore, the combination of IoT and big data allows for predictive capabilities in cognitive robotics. By constantly collecting and analyzing data, cognitive robots can anticipate potential issues or changes in their environment [34]. For example, a robot working in a manufacturing setting can use data from sensors to monitor equipment performance and identify signs of potential failures. With the help of big data analytics, the robot can predict when a machine is likely to malfunction and take proactive measures such as scheduling maintenance or alerting the human operator. Overall, IoT and big data play a crucial role in enabling cognitive robotics to operate intelligently and autonomously. They provide the necessary data input for the robots to perceive and understand their environment, make informed decisions, learn from past experiences, and predict future outcomes. In conclusion, IoT and big data are essential components in the field of cognitive robotics as they enable the collection, analysis, and utilization of massive amounts of data.

8.4 CHALLENGES, LIMITATIONS AND FUTURE DIRECTIONS

Cognitive robotics, as a branch of AI, has shown great potential in revolutionizing the industrial sector. However, there are still several challenges and future directions that need to be addressed to fully realize the benefits of cognitive robotics for Industry 5.0 [35]. One of the major challenges in cognitive robotics for Industry 5.0 is the development of robots with human-level intelligence. Robots with human-like dexterity and mobility for common tasks are still a challenge to create. Another challenge is the integration of cognitive abilities, such as perception, learning, and decision-making, into robots. This requires advancements in computer vision and deep learning techniques to enable robots to process and understand complex visual information in unstructured environments. Additionally, ensuring safe and efficient human-robot collaboration is a key challenge. This involves guaranteeing the safety of humans working close to robots, as well as designing intuitive interfaces that allow easy programming and interaction between humans and machines.

Handling exponential data growth, technology integration, workforce and training, financial and investment issues, cyber security, operational and implementation risks, and information and knowledge gaps are some of the difficulties facing cognitive robotics. Industry 5.0 research trends for cognitive robotics include mass customization, intelligent systems, collaborative robots (co-bots), and human-machine coexistence and connectivity. In Industry 5.0, future directions for cognitive robotics include creating efficient systems for human-robot collaboration, mitigating the danger of information and knowledge gaps, guaranteeing cyber security, and advancing social and societal advantages. By giving human operators access to robots' technical accuracy and heavy-lifting capabilities, Industry 5.0 seeks to restore the human element to development and production. It also permits a high degree of control and individualization of production phases. The development of collaborative robots that can interact with people intuitively will enable the application of digital twin technology in the industry.

In Industry 5.0, human-robot collaboration systems will play a bigger role. This means that efficient mechanical systems and collision detection procedures will be necessary. Businesses will also need to handle data efficiently for scalable AI solutions, which will require a robust data environment to handle high computational demands. Industry 5.0 emphasizes the use of highly qualified workers and robots to produce customized goods, such as cars and smart devices, while simultaneously advancing social benefits and sustainability. Since smart cities include smart houses, which include IoT-enabled heating, security, and other systems that communicate with one another to improve comfort and save energy, they are seen as significant IoT applications. Mass customization along with the utilization of computers, digital logic, integrated circuit chips, digital mobile phones, and the internet are the main components of Industry 5.0. Technological breakthroughs are digitalizing technology and changing both traditional products and corporate processes. Industry 4.0 is the marriage of cutting-edge technologies like cloud computing, IoT, cyber-physical systems, and AI with physical assets. To summarize, Industry 5.0's cognitive robotics seeks to handle data management, cyber security, social and societal hazards, and human-robot collaboration while also mass customizing and promoting sustainability. Prospects for research include developing efficient systems for human-robot collaboration and advancing social and societal advantages. Current research trends include integrating collaborative robots, intelligent systems, and human-machine connectivity.

Cognitive robots demand significant investment in research, hardware, and software. Some producers may find the expense of this complex machinery, combined with the need for specialized infrastructure and qualified workers, to be prohibitively expensive. The reliance on massive volumes of data for AI and ML creates issues of data security and privacy. Robust cyber security measures are required to safeguard sensitive data from breaches and promote appropriate data management practices. Cognitive robotics, like other forms of automation, has the potential to disrupt the workforce. As robots take on additional responsibilities, certain jobs may be replaced. However, Industry 5.0 is also projected to open new jobs in fields such as robot maintenance, programming, and data analytics. As robots become more autonomous, it is important to carefully analyze their ethical consequences. Questions are raised concerning robot decision-making, algorithmic biases, and the overall influence of intelligent robots on society. Clear ethical rules for the development and deployment of cognitive robots are essential.

8.5 CONCLUSION

A revolutionary era in manufacturing has begun with the introduction of Industry 5.0, which is defined by the smooth coexistence of sophisticated robotic technology and human labor. Cognitive robots, intelligent machines equipped with the ability to perceive, reason, and learn, can potentially transform many industries, from healthcare to manufacturing. They make proactive decisions and independently reconfigure their functions according to the situation, optimizing production schedules, minimizing downtime, and ensuring quality control. The potential for cognitive robots to transform a variety of industries and improve human-machine collaboration is seen in their extensive range of applications, including senior care, retail, city, and government services, medical and healthcare, personal assistants, automobiles, and more. Cognitive robots play a multifaceted role in Society 5.0, driving innovation, enhancing human capabilities, and fostering sustainable development across various sectors. IoT and big data play crucial roles in enabling cognitive robotics to function effectively. IoT provides the foundation for gathering massive amounts of data from various sensors and devices, which are then transmitted to the cloud for storage, analysis, and processing. Big data technologies enable the processing and analysis of large volumes of data in real-time, providing valuable insights into the robot's environment, and allowing it to make informed decisions and perform complex tasks. In Industry 5.0, human-robot collaboration systems will play a bigger role, necessitating efficient mechanical systems and collision detection procedures. Businesses will also need to handle data efficiently for scalable AI solutions, requiring a robust data environment to handle high computational demands. Industry 5.0 emphasizes the use of highly qualified workers and robots to produce customized goods while advancing social benefits and sustainability. Cognitive robots demand significant investment in research, hardware, and software, and may disrupt the workforce. However, it is projected to open new jobs in fields such as robot maintenance, programming, and data analytics. Clear ethical rules for the development and deployment of cognitive robots are essential.

REFERENCES

[1] K. Kawamura, and W. Browne, Cognitive Robotics. In: Meyers, R. (eds) *Encyclopedia of Complexity and Systems Science*. Springer, New York, NY, 2009. https://doi.org/10.1007/978-0-387-30440-3.

[2] O. Russakovsky, J. Deng, H. Su, J. Krause, S. Satheesh, S. and Ma, et al. Imagenet large-scale visual recognition challenge. *International Journal of Computer Vision*, vol. 115, 2015, pp. 211–252. https://doi.org/10.1007/s11263-015-0816-y

[3] I. Kotseruba, and J. K. Tsotsos, 40 years of cognitive architectures: core cognitive abilities and practical applications, Artificial Intelligence Review, vol. 53, 2020, pp. 17–94. https://doi.org/10.1007/s10462-018-9646-y

[4] Jinseok Woo, János Botzheim, and Naoyuki Kubota, System integration for cognitive model of a robot partner, *Intelligent Automation & Soft Computing*, 2017, 1-14. https://doi.org/10.1080/10798587.2017.1364919

[5] I. Giorgi, A Cangelosi, and G. L. Masala, Learning actions from natural language instructions using an ON-world embodied cognitive architecture, *Frontiers in Neurorobotics*, vol. 15, 2021, 626380. https://doi.org/10.3389/fnbot.2021.626380

[6] R. M. Damgaard, R. Pedersen and T. Bak, Toward an idiomatic framework for cognitive robotics. *Patterns*, vol. 3, 2022. 100533. July 8, 2022 https://doi.org/10.1016/j.patter.2022.100533

[7] K. Miyazawa, T. Horii, T. Aoki and T. Nagai, Integrated cognitive architecture for robot learning of action and language. *Frontiers in Robotics and AI*, vol. 6, 2019, p. 131. https://doi.org/10.3389/frobt.2019.00131

[8] I. Andras, E. Mazzone, F.W. van Leeuwen, G. De Naeyer, M.N. van Oosterom, S. Beato, T. Buckle, S. O'Sullivan, P.J. van Leeuwen, and A. Beulens, Artificial intelligence and robotics: A combination that is changing the operating room, *World Journal of Urology*, vol. 38, 2020, pp. 2359–2366.

[9] A. Verma and S. Kumar, "Cognitive Robotics in Artificial Intelligence," *2018 8th International Conference on Cloud Computing, Data Science & Engineering (Confluence)*, Noida, India, 2018, pp. 65–70. https://doi.org/10.1109/CONFLUENCE.2018.8442725

[10] Mohsen Soori, Behrooz Arezoo, and Roza Dastres, Artificial intelligence, machine learning and deep learning in advanced robotics, a review, *Cognitive Robotics*, vol. 3, 2023, pp, 54–70, ISSN 2667-2413, https://doi.org/10.1016/j.cogr.2023.04.001

[11] Q. Bai, S. Li, J. Yang, Q. Song, Z. and Li, X. Zhang, Object detection recognition and robot grasping based on machine learning: A survey, *IEEE Access*, vol. 8, 2020, pp. 181855–181879.

[12] J. Howard, Artificial intelligence: Implications for the future of work, *American Journal of Industrial Medicine*, vol. 62, 2019, pp. 917–926.

[13] Evi Zouganeli and Athanasios Lentzas. 2023. Artificial cognitive functions towards AI-enabled collaborative robots. In *Proceedings of the 16th International Conference on PErvasive Technologies Related to Assistive Environments (PETRA '23)*. Association for Computing Machinery, New York, NY, USA, 14–16. https://doi.org/10.1145/3594806.3594836

[14] W.J. Lee, H. Wu, H. Yun, H. Kim, M.B. and Jun, J.W. Sutherland predictive maintenance of machine tool systems using artificial intelligence techniques applied to machine condition data, *Procedia Cirp*, vol. 80, 2019, pp. 506–511.

[15] Francesco Semeraro, Alexander Griffiths, and Angelo Cangelosi, Human-robot collaboration and machine learning: A systematic review of recent research, *Robotics and Computer-Integrated Manufacturing*, vol. 79, 2023, 102432, ISSN 0736-5845, https://doi.org/10.1016/j.rcim.2022.102432

[16] Harley Oliff, Ying Liu, Maneesh Kumar, Michael Williams, and Michael Ryan, Reinforcement learning for facilitating human-robot-interaction in manufacturing, *Journal of Manufacturing Systems*, vol. 56, 2020, pp. 326–340, ISSN 0278-6125, https://doi.org/10.1016/j.jmsy.2020.06.018

[17] Xing Li, Junpei Zhong, and M.M. Kamruzzaman, Complicated robot activity recognition by quality-aware deep reinforcement learning, *Future Generation Computer Systems*, vol. 117, 2021, pp. 480–485, ISSN 0167-739X, https://doi.org/10.1016/j.future.2020.11.017

[18] Séverin Lemaignan, Mathieu Warnier, E. Akin Sisbot, and Aurélie Clodic, Rachid Alami, Artificial cognition for social human–robot interaction: An implementation, *Artificial Intelligence*, vol. 247, 2017, pp. 45–69, ISSN 0004-3702, https://doi.org/10.1016/j.artint.2016.07.002

[19] Debasmita Mukherjee, Kashish Gupta, Li Hsin Chang, and Homayoun Najjaran, A survey of robot learning strategies for human-robot collaboration in industrial settings, *Robotics, and Computer-Integrated Manufacturing*, vol. 73, 2022, p. 102231, ISSN 0736-5845, https://doi.org/10.1016/j.rcim.2021.102231

[20] Thomas Tawiah, Machine learning and cognitive robotics: opportunities and challenges, *Cognitive Robotics and Adaptive Behaviors. IntechOpen*, 2022. https://doi.org/10.5772/intechopen.107147

[21] J. Hertzberg and A. Saffiotti, Using semantic knowledge in robotics, *Robotics and Autonomous Systems*, vol. 56, no. 11, 2008, pp. 875–877.

[22] G. Randelli, *Improving Human-Robot Awareness through Semantic-driven Tangible Interaction*, Ph.D. Thesis, Sapienza University of Rome, 2011.

[23] J. Marciniak and Z. Vetulani, Ontology of spatial concepts in a natural language interface for a mobile robot, *Applied Intelligence*, vol. 17, no. 3, 2002, pp. 271–274.

[24] M. Skubic, C. Bailey, and G. Chronis, A sketch interface for mobile robots, in *Proceedings of the IEEE International Conference on Systems, Man and Cybernetics*, 2003, pp. 918–924.

[25] J.M. Kruijff, H. Zender, P. Jensfelt, and H.I. Christensen, Situated dialogue and spatial organization: What, where... and why, *International Journal of Advanced Robotic Systems*, vol. 4, no. 2, 2007, pp. 125–138.

[26] M. Hasanuzzaman, T. Zhang, V. Ampornaramveth, H. Gotoda, Y. Shirai, and H. Ueno, Adaptive visual gesture recognition for human-robot interaction using a knowledge-based software platform, *Robotics and Autonomous Systems*, vol. 55, no. 8, 2007, pp. 643–657.

[27] J. Modayil and B. Kuipers, Bootstrap learning for object discovery, in *Proceedings. of the IEEE/RSJ International Conference on Intelligent Robots and Systems*, 2004, pp. 742–747.

[28] T. Taniguchi, D. Mochihashi, T. Nagai, S. Uchida, N. Inoue, I. Kobayashi, and T. Inamura, Survey on frontiers of language and robotics, *Advanced Robotics*, vol. 33, no. 15–16, 2019, pp. 700–730. https://doi.org/10.1080/01691864.2019.1632223

[29] Y. Kilicaslan, and G. Tuna, An NLP-based approach for improving human-robot interaction, *Journal of Artificial Intelligence and Soft Computing Research*, vol. 3, no. 3, 2013, pp. 189–200.

[30] H.A. Younis, N.I.R. Ruhaiyem, W. Ghaban, N.A. Gazem, and M. Nasser, A systematic literature review on the applications of robots and natural language processing in education, *Electronics,* vol. 12, 2023, p. 2864. https://doi.org/10.3390/electronics12132864

[31] E González-Sarmiento et al., IOP conference series, *Materials Science and Engineering,* vol. 1154. p. 012008.

[32] Q. Wu et al., Cognitive internet of things: A new paradigm beyond connection, *IEEE Internet of Things Journal*, vol. 1, no. 2, 2014, pp. 129–143, April.https://doi.org/10.1109/JIOT.2014.2311513

[33] A. Satyanarayana, J. Kusyk, and Y. -W. Chen, Design of Cloud Based Robots Using Big Data Analytics and Neuromorphic Computing, *Proceedings of the 2018 IEEE Canadian Conference on Electrical & Computer Engineering (CCECE)*, Quebec, QC, Canada, 2018, pp. 1–4. https://doi.org/10.1109/CCECE.2018.8447874

[34] M. Mazhar Rathore, Awais Ahmad, Anand Paul, and Seungmin Rho, Urban planning and building smart cities based on the internet of things using big data analytics, *Computer Networks*, vol. 101, 2016, pp. 63–80, ISSN 1389–1286. https://doi.org/10.1016/j.comnet.2015.12.023

[35] Gissel Velarde, Artificial Intelligence and its Impact on the fourth industrial revolution: A review, *International Journal of Artificial Intelligence & Applications (IJAIA)*, vol 10, pp 41-48, no.6, November 2019.

Chapter 9

Industry 5.0 Unveiled, Precision Agriculture Empowered

Integrating Recommendation and Prediction Systems for Transparent Farming Transactions

Kaustubh Rathod, Devesh Rathi, and Sankalp Naranje

9.1 INTRODUCTION

Precision agriculture is a farming approach that utilizes technology, data analysis, and automation to optimize agricultural practices. It aims to maximize crop yields, minimize resource waste, and improve overall efficiency in the agricultural industry. One of the key components of precision agriculture is the integration of recommendation systems and predictive analytics. Algorithms in recommendation systems recommend the most pertinent and helpful products to users. Recommendation systems in the context of precision agriculture offer farmers tailored advice on a range of farming-related topics. Based on historical and current data, these recommendations may include the best dates to plant certain crops, the kinds of crops to use, fertilizers, pesticides, and irrigation techniques.

Recommendation systems rely on data sources like weather, soil quality, past crop performance, and sensor data from the farm to produce appropriate suggestions. These systems analyze and interpret the data using machine learning techniques, taking into account a variety of aspects like farmer preferences, crop requirements, and environmental circumstances. Farmers are able to make well-informed decisions regarding their farming techniques by incorporating recommendation systems into precision agriculture. Based on the system's customized recommendations, they may increase crop output, minimize waste, and allocate resources optimally.

A further essential element of precision agriculture is predictive analytics. It entails utilizing both past and current data to forecast upcoming occurrences or results. Predictive analytics can assist farmers in identifying possible obstacles or opportunities in precision agriculture. Predictive analytics models can anticipate future results by evaluating past data on market trends, pest and disease outbreaks, weather patterns, and agricultural performance. These models can, for instance, forecast when crops should be harvested, spot possible illnesses or pest infestations, or gauge consumer demand for particular products. Precision agricultural predictive

DOI: 10.1201/9781003501220-9

analytics models employ a variety of methods, including data mining, machine learning algorithms, and statistical analysis. They find patterns in big databases, analyze them, and produce insights that help farmers make wise decisions.

There are many advantages to precision agriculture when recommendation systems and predictive analytics are combined. Farmers can increase the efficacy and efficiency of their farming operations by merging recommendation systems' personalized advice with the insights produced by predictive analytics models. For instance, if a recommendation system receives information from a predictive analytics model indicating a possible pest epidemic, it can notify the system and offer the best pesticide to use along with the best way to apply it. Similar to this, a recommendation system can advise farmers on the best irrigation technique to lessen the impact of a meteorological event that a predictive model predicts may have an influence on the crops.

Precision agriculture is able to adjust to changing conditions in real time because to the integration of recommendation systems and predictive analytics. Farmers can improve the quality of their decision-making process and attain better results by regularly evaluating data, revising recommendations, and sending out timely notifications.

9.2 LITERATURE SURVEY

The body of research on precision agriculture's data-driven methods emphasizes how important it is to use trustworthy data sources to improve farmers' ability to make predictions and make decisions. According to research, weather stations are essential for delivering consistent and reliable weather data, which improves the accuracy of predictive models. Farmers are given practical insights by the incorporation of weather data, which helps forecast crop growth patterns and disease outbreaks.

The welfare of farmers and how it relates to energy efficiency and farm sustainability has garnered attention from scientists around the world in recent decades. This study looks at how Agriculture 5.0 helped farmers become prosperous after the epidemic and how they gradually changed their operations to become energy-smart farms. When combined with alternative energy sources, emerging technologies included in Agriculture 5.0 can offer affordable access to financing, weather updates, remote monitoring, and future energy solutions for the creation of smart farms. Numerous changes have been made to the production, distribution, and consumption of agricultural products by the authors in [1]. Under this environment, the COVID-19 pandemic has emerged as yet another significant modern threat to the agriculture industry. The notion of precision farming, sometimes known as smart farming, is crucial for tackling the present issues in agriculture related to food safety, sustainability, productivity, and climate change. It is insufficient to merely raise the procedure used to produce agricultural goods. Achieving and maintaining a high nutritional content while maintaining the required level of safety is equally crucial. Therefore, it becomes crucial to construct agricultural ecosystems based on developing technology in order to handle all of the aforementioned problems. This essay emphasized the value and significance of smart farming in the context of a resilient and sustainable agricultural ecosystem. The article also discussed the shift from smart farming to the new "digital era" of Agriculture 5.0 and the development of the energy-smart farm as a new agricultural paradigm. Agricultural commodity price variations have an impact on a nation's gross domestic product (GDP). Estimating and evaluating crop prices helps you make a wise choice before planting a particular kind of crop. Knowing how much a crop will cost will enable you to make smarter decisions, which will reduce losses and manage the risk of price changes. In [2], we used historical rainfall and Wholesale Price Index (WPI) data analysis to forecast the price of several crops. We analyzed the historical data, predicted the price for the most recent data, and estimated the price for the upcoming twelve months using the decision tree regressor (supervised machine learning method).

Using efficient machine learning methods and technologies, the data is shown on a Flask webpage with an intuitive user interface. The obtained training datasets offer sufficient information to predict the necessary pricing and market demand. As a result, the program enables farmers to earn more money and deal with less issues. Crop price prediction can be accomplished with a variety of algorithms, including deep learning, neural networks, decision trees, and support vector machines. Decision tree regressor, a supervised machine learning technique, was employed in our model. To improve accuracy, it is trained on a variety of Kharif and Ragi crops, such as barley, wheat, cotton, and paddy. The focus of current research is now on estimating plant physiological features rather than structural factors thanks to these technologies. There is discussion about futuristic methods to profit from these state-of-the-art technologies [3]. Researchers, scholars, and young people hoping to contribute to the realization of sustainable agriculture can benefit from this study. These efforts also hold the promise of a sustainable Agriculture 5.0. To demonstrate the significance of Remote Sensing (RS) technology in accomplishing Sustainable Development Goal 2 (SDG-2), the study sought to provide an overview of the key players in this new era of the 5th industrial revolution (a.k.a. Agriculture 5.0), including satellites, Robotic Process Automations (RPAs), and sensors. RS technology relies heavily on satellites, which can be enhanced with Internet of Drones (IoD) to fully realize its promise for ensuring sustainable agriculture. The current state of advanced farm management systems is reviewed in [4] by going over every important step again, from variable rate applications to data collection in crop fields. This allows growers to make the best decisions possible to save costs, protect the environment, and change how food will be produced to sustainably meet the impending population growth. This analysis validates that best decisions are made when one has consistent knowledge regarding farms. Farm data can be managed by agricultural management systems so that outcomes are coordinated to meet specific needs for every farm. Together with robotics and artificial intelligence (AI), this digital assistance for farmers ushers in the future of agriculture.

Turnips, muskmelons, kailan, peanuts, cantaloupes, and water mimosas were the six alternative crop kinds that were included in the price data for the monthly time series analysis of Hom Pathum Rice that was conducted in reference [5]. Given their restricted access to information technology systems, common farmers may find great value in the research findings. Farmers and agricultural communities can analyze the prediction methods without governmental backing or state-of-the-art equipment because they are straightforward. Thai farmers will benefit from this cutting-edge research by being able to apply smart farming and increase their profits. The decision tree algorithm, a supervised machine learning technique, is implemented by the authors in [8]. The suggested model was created to assist farmers in selecting the crop that will be most productive at the place and time of their choice for sowing. Our technology provides the farmer with helpful information well in advance of the cultivation process by forecasting the crop of choice's yield and price. Crop yield and price prediction can be done using a variety of prediction algorithms, including support vector machines (SVM), neural networks, and decision trees. Decision trees are used in our model. It has received good accuracy training on a variety of kharif and ragi crops, including barley, bhajra, arhar, and rice. In [9], the method determines the optimal crop for cultivation by taking into account a number of factors, including the weather forecast and the state of the soil. Utilizing the crop predictor can help reduce losses in the event of unfavorable circumstances. When ideal growth circumstances are possible, farmers can employ this technology to increase agricultural yield rates. In order to estimate the price for the next twelve months based on the Minimum Support Price (M.S.P.) and prices from the previous year or month, the government determines the M.S.P., which is taken into account by the system for forecasting crop prices. The farmer can compute income thanks to it.

Data analytics methods were used in [10] to estimate agricultural price analysis using the available data. In order to extract valuable information from the agricultural datasets, linear regression is used. A neural network is built for price prediction in order to boost the accuracy rate. To correctly measure the accuracy of each system used, the root mean square error is determined for each technique, and the most accurate system is then chosen. Researchers compute each of these parameters independently in [14] using machine learning methods and Internet of Things-based (IoT-based) devices. In terms of soil fertility, soil moisture, pH levels, and other variables, sensors are giving precise results. They can be more automated, precise, and economical by integrating them with image processing. Currently, farmers are utilizing their own estimates to calculate all of these characteristics, which is not producing very profitable results. In [16], time-series and machine learning algorithms are used to forecast the monthly prices of arecanut in Kerala. The root mean square error (RMSE) value on the arecanut dataset with prices from 2007 to 2017 was used to assess the performance of the models Seasonal AutoRegressive Integrated Moving Average (SARIMA), Long Short-Term Memory (LSTM) neural network, and Holt-Winter's Seasonal approach. It is difficult to make timely decisions based on intuition when commodity prices are volatile. On the basis of this work, decisions that support a better economic outcome can be taken. Kerala is the second-largest arecanut producer in the world, therefore the effort will assist arecanut growers in better managing their resources to produce more crops at the highest possible profit margin based on pricing forecasts.

9.3 METHODOLOGY

In this work, we take a thorough and iterative method to study recommendation and prediction system integration in agriculture. The goal of the research is to comprehend how these innovative technologies affect farmers' and sellers' decision-making processes, with an emphasis on crop selection, pricing tactics, and general farm management.

9.3.1 Objectives

Development of a novel online platform:

- With success, we have developed a cutting-edge web platform that enables smooth communication between farmers and retailers.
- We have created separate vendor and farmer logins with customized features to improve the user experience overall.

Integration of recommendation systems:

- Recommendation algorithms were successfully included into the platform to give farmers access to data-driven insights.
- investigated how recommendation technologies affect the chain of agriculture's decision-making processes.

Implementation of prediction algorithms:

- The tool incorporates sophisticated prediction algorithms to aid suppliers in making strategic decisions.

- We have taken into account variables including profitability, demand patterns, and market circumstances to improve forecast accuracy.

Privacy-conscious strategy:

- Developed and put into practice a strong, privacy-aware approach for building farming recommendation and prediction systems.
- We have ensured that data was handled and stored securely, placing special emphasis on protecting sensitive transactional data.

User testing and feedback:

- Carried out a thorough user test to evaluate the web application's usability and efficacy.
- We have gathered insightful input from vendors and farmers, which helped to improve and refine the platform.

Multilingual support and accessibility:

- Successfully added language support to the program for both Marathi and English.
- Assessed how easily the platform could be used by users with different levels of technology expertise.

Data analysis and dashboard creation:

- Evaluated the application's influence on farming transactions, efficiency, and transparency by analyzing transactional data it produced.
- Created an extensive dashboard that promotes accountability and transparency by showing historical transactional data.

Evaluation of impact on agricultural transactions:

- Carried out a comprehensive analysis of the integrated recommendation and prediction systems' overall influence on the reorganization of agricultural transactions.
- Assessed how much the application helps farmers and suppliers achieve financial sustainability, equity, and efficacy.

9.3.2 Proposed Approach

We foresee a complete plan for our research method, starting with a detailed needs evaluation of the agricultural industry. Through a comprehensive examination, our goal is to pinpoint particular obstacles and areas of discomfort in the decision-making procedures of farmers and vendors. We will concentrate on matters like crop selection, pricing tactics, and overall farm administration. Next, we will do a thorough assessment of the literature, looking at previous research on Industry 5.0, precision farming, and the incorporation of recommendation and prediction systems into farming transactions. This examination of the literature will act as a basis, enabling us to discover areas that could use improvement and gaps in our existing understanding.

A fundamental component of our suggested methodology entails proactive cooperation with entities within the agricultural ecosystem. We will interact with farmers, suppliers, and other pertinent parties to learn about their viewpoints, worries, and expectations. This cooperative effort will guarantee that the final product is in line with the realistic needs of the industry in addition to providing valuable insights for the growth of our web platform. Our goal is to develop a solution that appeals to the various parties involved in agricultural transactions by promoting an inclusive approach.

We will next start working on the design and implementation of an innovative web platform, building on this basis. With customized features to improve the user experience, this platform would include separate logins for farmers and vendors. A major area of concentration will be the integration of prediction and recommendation systems, which will enable suppliers to make strategic decisions and provide farmers with data-driven insights. The development process will be guided by a privacy-conscious strategy that ensures secure data handling and storage to protect critical transactional information.

In-depth user testing with input from farmers and vendors will be carried out to confirm the platform's efficacy and usability. To serve a varied user base, multilingual support for the Marathi and English languages will be put into place. Its effects on farming transactions, efficiency, and transparency will become clearer through the ensuing study of transactional data produced by the platform. The development of an all-inclusive dashboard that showcases historical transactional data will additionally enhance transparency and accountability in the agricultural supply chain.

Finally, a thorough assessment of the integrated recommendation and prediction systems' overall influence on the restructuring of agricultural transactions is included in our suggested methodology. This assessment will determine how much the application advances efficiency, fairness, and long-term financial viability for all stakeholders, providing a comprehensive view of Industry 5.0's disruptive potential in precision agriculture.

9.3.3 Data Collection and Analysis for Farming Predictions

Accurately predicting agricultural results is essential for making well-informed farming decisions. It aids farmers in risk mitigation, productivity growth, and resource optimization. But precise forecasts necessitate a thorough approach to data gathering and analysis. This topic examines the essential components of gathering and analyzing data for farming forecasts, emphasizing the value of dependable data sources, data pretreatment strategies, and diverse analytical approaches.

- **Reliable data sources:** It is critical to obtain data from trustworthy sources in order to guarantee accurate forecasts in farming. Conventional techniques, such as soil samples, weather monitoring, and handwritten records, are still essential. But new developments in technology have opened up more options. Farmers are now equipped with dynamic insights thanks to real-time data from internet-connected sensors, satellite imaging, and remote sensing technologies. Sensor-equipped weather stations provide accurate and up-to-date weather information at particular places. Important data on pH, moisture content, and nutrient composition are obtained by soil sampling, which helps determine the best course for fertilization and irrigation. In addition, farmers can monitor crop health, identify pests and diseases, and make timely decisions based on extensive, high-resolution data thanks to remote sensing technology like drones and satellites and Internet of Things (IoT) devices like moisture sensors.
- **Data preprocessing techniques:** Correct preprocessing of gathered data is necessary prior to performing predictive analysis. By handling missing values, removing outliers, and ensuring data quality, these methods improve prediction accuracy. Finding errors, eliminating duplication, and dealing with inconsistent or missing data using methods like averaging and imputation are all part of data cleaning. Standardizing various data kinds and scales through normalization approaches like Z-score normalization and Min-Max scaling ensures comparability between variables. By removing superfluous characteristics and decreasing the dimensionality of the dataset, feature selection increases computational efficiency. Using statistical techniques and clustering algorithms, outlier detection finds and manages outliers that may have an effect on predictive models.
- **Analysis methods:** Depending on the objectives and data at hand, a variety of analysis techniques are available to forecast farming outcomes based on the gathered and preprocessed data. Regression analysis finds correlations between independent factors (like weather conditions and soil qualities) and dependent variables (like crop production) by using methods like linear and multiple regression. Temporal data, such daily weather or historical crop yields, require time series analysis using techniques like AutoRegressive Integrated Moving Average (ARIMA) and Seasonal and Trend Decomposition

using Loess (STL). Decision trees and support vector machines are two examples of machine learning algorithms that use past data to identify patterns. These algorithms are especially good at managing complicated relationships and big datasets. By taking into account the geographical linkages between data points, spatial analysis techniques like kriging and spatial regression models can forecast variabilities across fields or regions.

Accurate forecasting for farming operations depends on the collaboration of trustworthy data sources, careful data pretreatment, and suitable analysis techniques. With the right information at their disposal, farmers can efficiently reduce risks, maximize productivity, and optimize agricultural processes.

9.4 RESULTS AND DISCUSSION

9.4.1 Benefits of Precision Agriculture in Improving Productivity and Sustainability

- **Improved resource efficiency:** By using resources like water, fertilizer, and pesticides more effectively, precision agriculture helps farmers minimize waste and lessen their influence on the environment. Farmers can minimize water usage and prevent over-irrigation by precisely measuring the quantity of water crops require and applying it only where needed, all thanks to soil condition analysis. In a similar vein, farmers can optimize nutrient uptake and minimize unnecessary application by mapping nutrient deficits in the soil and applying fertilizers to specific regions.
- **Enhanced crop yield:** The capacity of precision agriculture to raise crop output is one of its main benefits. Farmers may obtain accurate real-time data on crop health, soil fertility, and moisture levels by utilizing cutting-edge technologies. Farmers are able to use this data to pinpoint variances in their fields and carry out site-specific interventions to deal with those variations. Farmers may optimize crop yield and quality and increase profitability by applying the appropriate quantity of inputs, such as irrigation, fertilizers, and pesticides, at the appropriate times and in the appropriate places.
- **Reduction of environmental impact:** Reducing the agricultural techniques' environmental impact is largely dependent on precision agriculture. Farmers can minimize the use of synthetic chemicals and help reduce contamination of water and soil by applying inputs only where and when needed. Applying pesticides with precision can also reduce chemical residues, reducing negative impacts on organisms that are not intended targets and fostering biodiversity. Accurately

determining the moisture content of the soil enables farmers to optimize irrigation techniques, saving water and energy.

- **Preserving soil health:** Degradation and erosion of soil pose serious problems for agriculture. Effective tools and methods for mitigating and managing these problems are provided by precision agriculture. Farmers may avoid over- or under-watering their crops, which can lead to erosion, by keeping an eye on the moisture content of their soil. Farmers can also apply soil conservation techniques, like contour ploughing or cover crops, in particular regions that have been identified as being vulnerable to erosion by evaluating the composition of the soil. All things considered, precision agriculture supports long-term sustainability and productivity by maintaining the health of the soil.
- **Data-driven decision making:** Gathering and analyzing enormous volumes of data is crucial to precision agriculture. Farmers are able to make more educated judgements about crop management techniques because to this data-driven strategy. Farmers can take preventative action by using remote sensing technologies to collect data on pest infestations, nutrient deficits, and crop health. This data may be processed by sophisticated analytics and machine learning algorithms, which can then offer predicted insights and help farmers take appropriate action. Farmers may maximize yield and raise overall productivity with the help of recommendations and real-time data access.

9.4.2 The Impact of Industry 5.0 on Agriculture

Industry 5.0 revolutionizes agriculture by addressing some of the industry's key challenges:

- **Increased efficiency and productivity:** The productivity and operational efficiency of agriculture are increased by the integration of Industry 5.0 technology. Robotic seed planting and harvesting are examples of automated processes that reduce labor needs and speed up production. AI-driven monitoring systems scan data continually for anomalies, allowing for the early identification of agricultural illnesses or irrigation problems. Higher yields and less production losses are the results of this proactive strategy.
- **Resource optimization:** Industry 5.0-driven precision agriculture maximizes the use of resources including pesticides, fertilizers, and water. Based on crop requirements and environmental parameters, farmers may accurately predict the quantity of resources needed with the use of AI algorithms. Industry 5.0 minimizes waste, lessens

its negative effects on the environment, and increases sustainability through resource optimization.

- **Risk reduction:** Recommendation and prediction systems used with agriculture lowers risk of disease outbreaks, variable crop yields, and changes in market demand. Farmers can avoid financial losses, make data-driven decisions, and proactively manage potential hazards when they have access to real-time data and AI-driven insights.
- **Sustainable agricultural practices:** Industry 5.0 minimizes dependency on hazardous chemicals and maximizes resource utilization, allowing farmers to embrace sustainable farming methods. By applying fertilizers and insecticides precisely, precision agricultural techniques reduce the negative environmental effects of these practices. Industry 5.0 supports sustainable farming methods by maintaining soil health and maximizing water use.
- **Increased profitability:** Agriculture may become more profitable because to Industry 5.0's operational optimization, waste reduction, and risk mitigation strategies. The incorporation of cutting-edge technologies raises crop yields, lowers expenses, and increases farm management in general. Farmers may enhance their earnings and sustain a competitive advantage in the market by utilizing big data analytics and AI to facilitate improved decision-making.

9.4.3 Krishi Vyapaar Sangam: The AgriTrade Connect

The implementation technique of "KrishiVyaparSangam: The AgriTrade Connect" commences with an extensive needs assessment conducted within the agricultural sector. The objective of this assessment is to identify obstacles and gather necessary requirements for farming transactions that are transparent. In order to make sure that the platform effectively meets the needs of the industry, farmers, vendors, and other pertinent entities actively participate in the initial phase of stakeholder engagement. The design and development of the platform are then guided by a survey of the literature on Industry 5.0, precision farming, and data-driven methodologies; the incorporation of new trends is informed by an examination of the state of technology today.

Different logins for farmers and sellers are part of the platform design, and features are customized based on feedback from users. A key component is the integration of prediction and recommendation systems, which provides consumers with data-driven insights for optimal decision-making. Reliable sources like weather stations, soil sampling networks, remote sensing technologies, and IoT devices are incorporated into the platform to guarantee data accuracy. Sensitive transactional data is secured using a strong privacy-aware approach that includes user authentication and encryption.

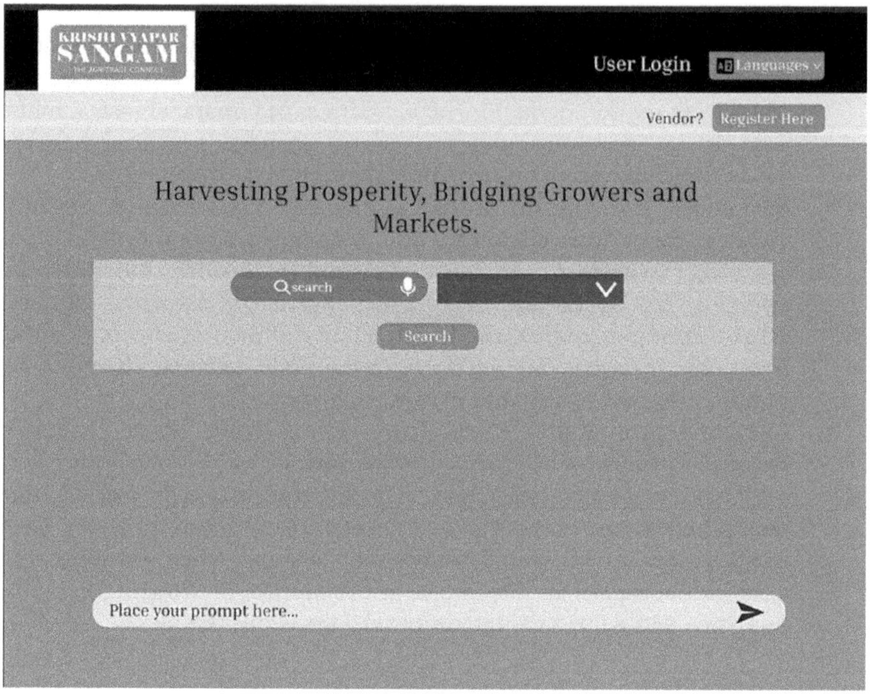

Figure 9.1 Landing/Home Page of Web-Based Application.

The use of the innovative web-based application has produced significant outcomes in the restructuring of farming transactions. Recommendation and prediction systems have been integrated, empowering farmers and vendors to make better decisions and increasing the agricultural supply chain's overall efficiency.

The platform has facilitated negotiated agreements between farmers and vendors, which have improved transaction fairness. Farmers profit when their food is sold at prices set by the market, while vendors use predictive algorithms to make strategic choices based on market conditions, profitability, and seasonal demand. The agricultural market is now more responsive and dynamic as a result of this mutually beneficial partnership.

Support for Marathi and Hindi is multilingual and takes into account cultural quirks and different farming methods in order to serve a wide range of users. Conducting user testing and gathering feedback is essential for evaluating usability, gaining understanding of user experiences, and pinpointing areas that require enhancement. Cleaning, normalization, feature selection, and outlier detection are a few examples of data preprocessing procedures that improve the quality and dependability of

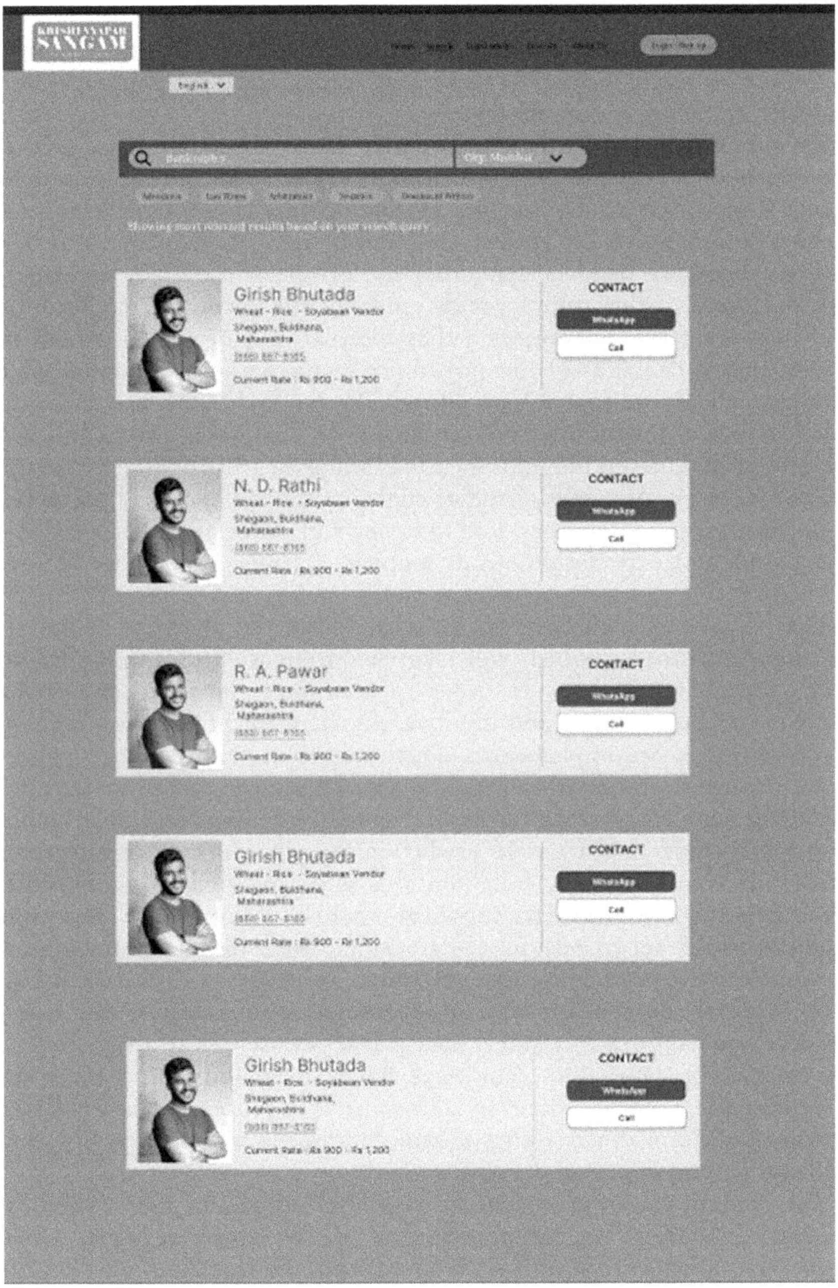

Figure 9.2 Landing/Home Page of Web-Based Application.

gathered data for predictive modelling. In addition to translation services, the multilingual assistance component included culturally appropriate design features. By taking into account regional accents, cultural quirks, and users' differing degrees of technology proficiency, the platform aimed to foster an inclusive community.

By presenting historical transactional data, thorough data analysis and dashboard design promote accountability and openness between farmers and vendors. During the evaluation phase, the overall effect of KrishiVyaparSangam on agricultural transactions is evaluated, and user insights regarding financial sustainability, equity, and efficacy are sought. The outcomes of this study provide guidance for upcoming modifications and improvements, guaranteeing that the platform develops to fulfil the ever-changing demands of the agriculture sector within the framework of Industry 5.0 and precision agriculture. The research team first evaluated the demands of the industry before using both qualitative and quantitative research techniques to pinpoint the needs and pain areas unique to the agriculture industry. Activities aimed at engaging stakeholders will encompass focus groups, questionnaires, and interviews in order to gather a range of viewpoints and expectations from suppliers, farmers, and other industry participants.

User experience will be given priority during the platform design and development process, which will take cues from well-functioning trading and e-commerce systems. In order to guarantee the platform's overall usability, responsiveness, and intuitiveness, iterative prototyping and user feedback loops are implemented. Furthermore, to improve user engagement, the incorporation of state-of-the-art technologies will be investigated, including augmented reality (AR) for virtual farm visits and three dimensional (3D) modeling for yield prediction. For the purpose of integrating trustworthy data sources, collaborations with government organizations and current agricultural data suppliers would be prioritized. Cooperation made it possible for correct, up-to-date data to enter the platform with ease, increasing its dependability and relevance. In order to ease data sharing with external sources, application programming interfaces (APIs) had to be developed during this stage. The privacy-conscious strategy involved the creation of transparent data usage standards in addition to encryption and authentication procedures to address privacy concerns. Users will be in charge of what information they provide, and the user interface will include tools for getting informed consent.

The continuous improvement strategy was utilized in user testing and feedback gathering. An enthusiastic staff of user testers regularly solicits input from a wide range of users, and incorporates this input into frequent updates and feature improvements. The platform's ability to adapt to shifting user needs and technology improvements is ensured by this iterative process. The construction of the dashboard and thorough data analysis went above

Figure 9.3 Landing/Home Page of Web-Based Application.

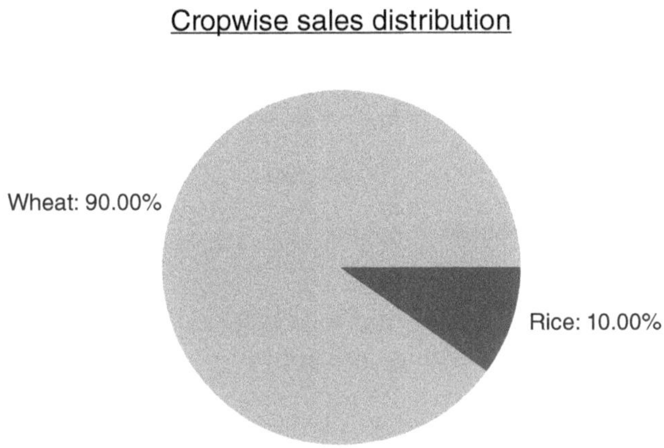

Figure 9.4 Cropwise Sales Distribution.

and beyond traditional reporting. Innovative data visualization methods are used to give farmers and suppliers useful information, such as trend analysis and geographic mapping. The dashboard assisted users in optimizing their farming practices by acting as a decision support system in addition to displaying transactional data. In addition to quantitative measures, user testimonials and case studies were used to gather qualitative data for the platform effect evaluation. The platform's ability to encourage efficiency, equity, and financial sustainability in agricultural transactions is measured by key performance indicators, or key performance indicator (KPIs).

Sensitive data security is ensured by the privacy-conscious methodology used in the development of the recommendation and prediction systems. The separate login credentials for farmers and suppliers, in conjunction with the all-inclusive dashboard, promote openness and responsibility, thereby

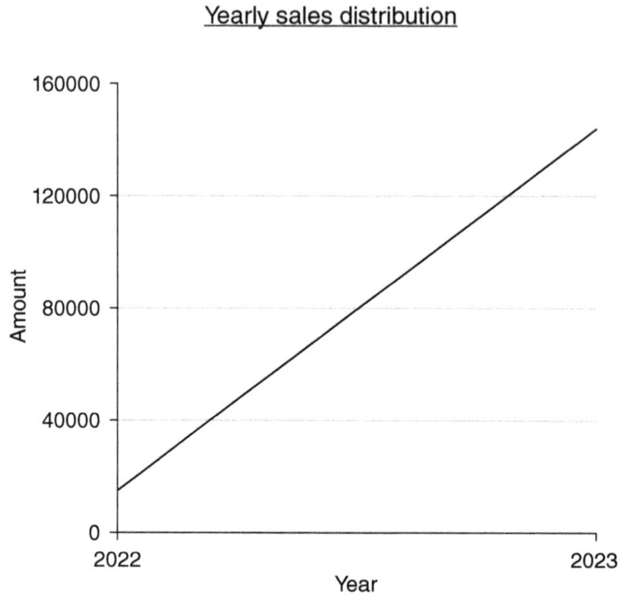

Figure 9.5 Yearly Sales Distribution.

cultivating mutual trust among the stakeholders. The platform's bilingual interface increases its accessibility and appeals to a broader range of users.

9.5 CONCLUSION

In order to guarantee that the established web platform is practically in line with industry needs, the suggested method placed a high priority on stakeholder participation. This involved interacting with farmers, vendors, and pertinent institutions. The innovative platform was created and put into use with the purpose of improving user experiences and enabling smooth interactions. It has unique logins, recommendation systems, and prediction algorithms. The utilization of data preprocessing methods, including feature selection, cleaning, normalization, and outlier detection, with the goal of improving the quality of the gathered data and raising prediction accuracy.

The study aims to create a revolutionary platform that not only tackles the present issues in precision agriculture but also establishes a standard for openness, effectiveness, and sustainability in agricultural dealings. The all-encompassing strategy, which includes technology integration, literature analysis, stakeholder involvement, and needs assessment, places the platform in a dynamic position within the rapidly changing Industry

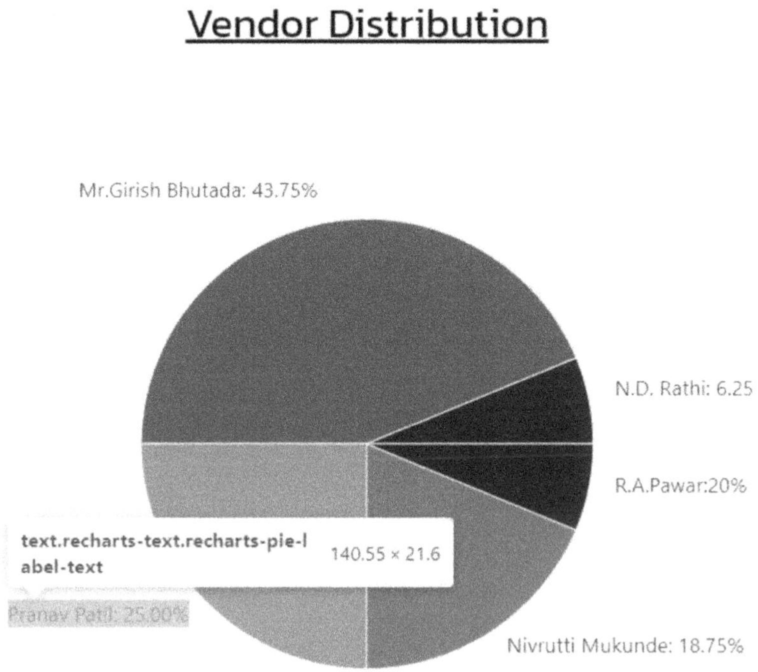

Figure 9.6 Vendor Distribution.

5.0 environment. By using user-centric design concepts that are guided by rigorous user testing and stakeholder involvement, KrishiVyaparSangam is able to maintain both technological advancement and intuitive accessibility for a wide range of users. The platform demonstrates a dedication to diversity by acknowledging the many language, cultural, and privacy preferences of the agricultural community through its multilingual support, cultural considerations, and privacy-conscious approach. The methodology takes into account the dynamic nature of technology and agriculture by anticipating continuous validation studies and future adjustments. The platform hopes to stay ahead of the curve and provide the agricultural sector with a sustainable and future-proof solution by continuing to be flexible enough to accommodate new technologies like quantum computing. The methodology that has been suggested for "KrishiVyaparSangam: The AgriTrade Connect" essentially goes beyond conventional limits by combining technical innovation, user experience, and stakeholder participation. As the project develops, it hopes to add to the scholarly conversation on precision agriculture as well as the real-world application of changing farming practices in the context of Industry 5.0.

Table 9.1 Comparison of Existing and Proposed Application

Feature	*Existing System*	*Proposed Application*
Focus	frequently concentrate on only one thing (e.g., yield prediction).	Combines forecasting and advice for a comprehensive strategy.
Data integration	Low data integration, which can necessitate manual data entry.	Incorporates transaction history, meteorological data, and market data.
Decision support	Insufficient backing for pricing and negotiating tactics.	Offers suggestions regarding crop choice, cost, and suppliers.
Transparency	Limited price and transaction transparency in the past.	Offers a comprehensive dashboard that includes transaction history.
User interface	Maybe not very user-friendly or multilingual.	Separate farmer and vendor interfaces that support both Marathi and English.
Privacy	Data gathering and usage may give rise to privacy problems.	Uses privacy-aware design for prediction and recommendation systems.

9.6 FUTURE SCOPE

The present study's examination of Industry 5.0's revolutionary potential in precision agriculture creates opportunities for additional research and development. The results and suggested methodology provide a solid foundation for a number of promising avenues for future research and development that have the potential to improve farming transactions' sustainability, efficiency, and transparency. Beyond the strategy that is now being used, this research will continue to provide a roadmap for future innovation and improvement in the field of precision agriculture. In the era of Industry 5.0, embracing these future prospects will help to improve farming transactions that are transparent, efficient, and sustainable as technology environments continue to change.

The methodology will incorporate regular validation studies and future adjustments as a forward-looking element. In order to further improve the platform's capabilities, this entails foreseeing upcoming technologies and investigating their possible integration, such as quantum computing for complicated data processing. To put it briefly, the extended methodology for "KrishiVyaparSangam: The AgriTrade Connect" strives for a sophisticated and all-encompassing approach, combining technological innovations, industry insights, and user-centric design principles to create a cutting-edge platform for transparent farming transactions and precision agriculture.

REFERENCES

[1] Konstantina Ragazou, Alexandros Garefalakis, Eleni Zafeiriou and Ioannis Passas, Agriculture 5.0: A New Strategic Management Mode for a Cut Cost and an Energy Efficient Agriculture Sector, *Energies* 2022, 15, 3113.

[2] Ranjani Dhanapal, A AjanRaj, S Balavinayagapragathish, and J Balaji, Crop price prediction using supervised machine learning algorithms, *Journal of Physics: Conference Series ICCCEBS* 2021, 1916, 1–9.

[3] Vanesa Martos, Ali Ahmad, Pedro Cartujo and Javier Ordoñez, Ensuring Agricultural Sustainability through Remote Sensing in the Era of Agriculture 5.0, *Applied Sciences* 2021, 11, 5911.

[4] Verónica Saiz-Rubio and Francisco Rovira-Más, From Smart Farming towards Agriculture 5.0: A Review on Crop Data Management, *Agronomy* 2020, 10, 207.

[5] Lakkana Ruekkasaem and Montalee Sasananan, Forecasting Agricultural Products Prices Using Time Series Methods For Crop Planning, *International Journal of Mechanical Engineering and Technology (IJMET)*, July 2018, 9 (7), 957–971, Article ID: IJMET_09_07_102.

[6] Thomas W. Hertel, Uris Lantz C. Baldos, and Dominique van der Mensbrugghe, Predicting Long-Term Food Demand, Cropland Use, and Prices, *Annual Review of Resource Economics* 2016, 8, 417–441.

[7] Paulo V. Cenas, Forecast of Agricultural Crop Price using Time Series and Kalman Filter Method, *Asia Pacific Journal of Multidisciplinary Research* November 2017, 5 (4), 15–21. Part III, P-ISSN 2350-7756, E-ISSN 2350-8442.

[8] Sadiq A. Mulla and Dr.S.A.Quadri, Crop-Yield and Price Forecasting Using Machine Learning, *The International Journal of Analytical and Experimental Modal Analysis* August/2020, XII (VIII), 731-1737. ISSN NO: 0886-9367.

[9] Ganesh Sastry Kakaraparthi, and B.V.A.N.S.S. Prabhakar Rao, Crop Price Prediction Using Machine Learning, *International Research Journal of Modernization in Engineering Technology and Science* June-2021, 03 (06), 3477–3481.

[10] Pandit Samuel, B.Sahithi, T.Saheli, D.Ramanika, and N.Anil Kumar, Crop Price Prediction System Using Machine learning Algorithms, *Quest Journals, Journal of Software Engineering and Simulation* 2020, 6 (1), 01–13. ISSN(Online):2321-3795 ISSN (Print):2321-3809.

[11] N Baryshnikova, P Altukhov, and N Naidenova, A Shkryabina, Ensuring Global Food Security: Transforming Approaches in the Context of Agriculture 5.0, *IOP Conference Series: Earth and Environmental Science (ISTC-EARTHSCI)*, 2022, 988, 1–6.

[12] Yuehjen E. Shao and Jun-Ting Dai, Integrated Feature Selection of ARIMA with Computational Intelligence Approaches for Food Crop Price Prediction, *Hindawi Complexity* 2018, 2018, Article ID 1910520, 1–17.

[13] V. Sellam and E. Poovammal, Prediction of Crop Yield using Regression Analysis, *Indian Journal of Science and Technology* October 2016, 9 (38), 1–5. 10.17485/ijst/2016/v9i38/91714.

[14] Swati Shirsath, Prof. Sanjay Pingat, Analyzing Cost of Crop Production and Forecasting the Price of a Crop, *International Research Journal of Engineering and Technology (IRJET)* May 2022, 09 (05), 198–202.

[15] Ashwini Darekar and A Amarender Reddy, Forecasting Wheat Prices in India, *Wheat and Barley Research*, 10 (1), 33–39.

[16] Kiran M. Sabu, and T. K. Manoj Kumar, Predictive Analytics in Agriculture: Forecasting prices of Arecanuts in Kerala, *Third International Conference on Computing and Network Communications (CoCoNet'19)*, 699–708.

Chapter 10

Cognitive Computing in Banking and Finance

Amar Buchade, Parikshit Mahalle, and Rajani Sajjan

10.1 INTRODUCTION TO COGNITIVE COMPUTING

The goal of cognitive computer research is to create artificial intelligence (AI) systems that can simulate human cognitive methods, such as perception, learning, goal setting, and reasoning. Cognitive systems will function at a pace and with an information capacity significantly faster than that of humans. These systems will collaborate with people on their terms and operate as a partner, advisor, helpmate, and co-creator. Cognitive computing offers a completely new paradigm for computing that can be applied to real-world issues with the help of parallel computers that communicate with humans and other cognitive systems. Cognitive systems will enable human-like reasoning, achieving capabilities that go beyond human capacity or are impractical for real-time human involvement. Cognitive computing emulates the processes of the human brain to analyze information and solve complex problems. Cognitive computing encompasses a range of disciplines and concepts, including neurobiology, reasoning, cognitive sciences, behavioral sciences, decision making, neural systems, cognitive informatics, human brain and big data, among others. AI enhances decision-making and reasoning processes, thereby promoting the advancement and application of cognitive computing.

Cognitive computing has many applications including:

- Healthcare: Assisting in healthcare diagnosis and recommending healthcare plans.
- Finance: Fraud detection and risk management.
- Customer service: Virtual assistants and chatbots.
- Education: Adaptive learning systems.
- Retail: Online shopping, personalized marketing, manufacturing predictive maintenance and cyber security, etc.
- Human resources: Talent acquisition and employee engagement, etc.

DOI: 10.1201/9781003501220-10

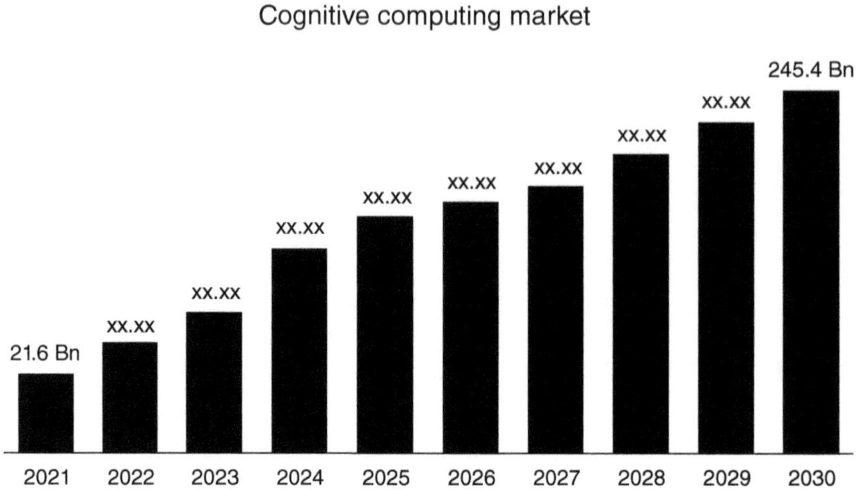

Figure 10.1 Growth of Cognitive Computing.

Banking and Finance are crucial to a country's economic stability and growth. Cognitive computing plays an important role in the financial and banking industry. [1–4].

The need for cognitive systems is increasing in large enterprises due to cloud-based services, and this trend is expected to spread to small and medium-sized businesses as well. Since cloud-based deployment reduces the cost of adopting cognitive computing within the company, there is an overall increase in demand for cloud-based services [5]. IBM Watson, a significant player in the cognitive computing business, intends to introduce Watson Analytics in future years. This cloud-based solution technology will hasten the cloud's adoption. The necessity of cognitive computing is growing quickly in industrialized nations and is anticipated to spread throughout the course of the forecast period. Figure 10.1 summarizes market insights from the Spherical Insights report, illustrating the projected growth of the cognitive computing market from USD 21.6 billion in 2021 to USD 245 billion by 2030 [6].

10.2 WORKING OF COGNITIVE BANKING

A cognitive system gathers a good deal of information in order to provide workers and consumers with guidance that is supported by facts. It is an autonomous learning system that imitates how the brain functions with natural language processing (NLP) and data mining capabilities. To process any decision, it receives input in the form of both internal and external

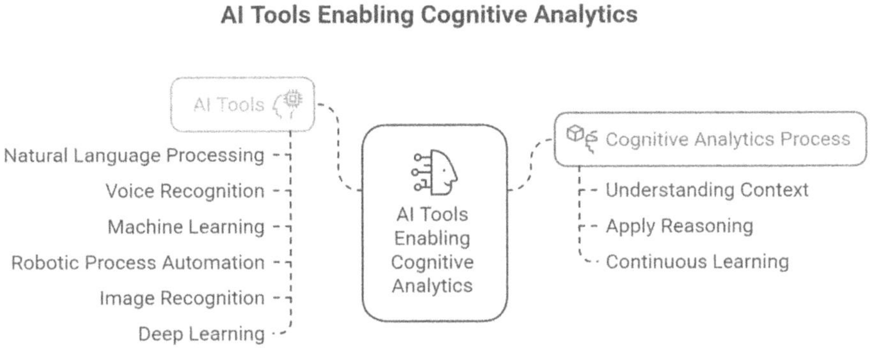

Figure 10.2 AI Tools enable Cognitive Analytics.

structured and unstructured data. Banks have accumulated unstructured data over the previous few decades and structured data from operational systems that are both extremely important to business. Unstructured data, which includes emails, photos, videos, audio, news, webpages, blogs, and social media, now comprises the majority of data that is accessible. A self-sustaining company environment is created by processing this raw material cognitively, allowing partners, employees, and consumers to communicate through an intuitive interface.

The automation of related processes in response to changing requirements is made possible through machine learning (ML), robotic process automation, NLP, smart workflows, and other cognitive tools. . By offering fresh perspectives and methods of operation, these technologies boost efficiency while decreasing the requirement for interpersonal contact. Unlike conventional computing models, which rely on pre-established programs and rules, cognitive systems are able to manage complex scenarios involving large amounts of data.

Figure 10.2 illustrates how AI tools drive cognitive analytics, enabling the understanding of context, the application of reasoning, and continuous learning. [1–4].

10.2.1 Advantages of Cognitive Banking

In the context of cognitive banking, there are several advantages and disadvantages that users should consider. The following section outlines some of these key factors.

- Agility and simplicity when consulting customers through the internet and with virtual assistants, at any time and from anywhere. Time is a critical factor for customers, and cognitive banking enhances the efficiency of numerous processes by significantly increasing speed.

- The integration and study of big data in banking enables the optimization of risks for both the bank and its clients. It allows for real-time reporting on customer segments and the status of their products, providing more personalized advice and facilitating improvements in a faster and simpler manner for managers. In summary, it delivers a more efficient and tailored service for both clients and the bank.
- The customer is the center of any business strategy. In this sense, the demands of the millennial generation in banking push us to digitize many processes and adapt to their needs. Millennials mainly seek: integration of services, agility and simplicity when making transactions and following procedures, immediacy in obtaining information and consulting where they have doubts, and a user-friendly personalized experience.

Cognitive banking responds to these interests.

10.2.2 Disadvantages of Cognitive Banking

- Identifying models and learning through the use of AI; for a field as delicate as the financial sector this process will require a lot of time and is still in a very incipient state to optimize results.
- The implantation of the technology in banking will reduce the number of employees, mainly in the branches. They will replace those repetitive processes that the employees do, giving them more time to dedicate to strategic tasks or advice. While this development is advantageous for customers, the resulting job displacement presents a significant concern, particularly in the context of the current global economic climate.
- The capture of new customers adapted to the use of technology is important, however, the retention of customers that have not been digitized remains equally important. There are still many clients who come to banking offices in need of a financial adviser and would prefer that be a human being. The large-scale implementation of this technology may lead to a loss of confidence among less digitally adept users, unless proper adaptation measures are in place.

10.3 NEED OF COGNITIVE COMPUTING IN BANKING SECTOR

Customers of financial services can receive personalized support through cognitive capabilities, which in turn positively influences their perception of the bank's operations. Cognitive computing enables banks to run more efficiently, manage risks, and provide better experiences for customers [7,

8]. Cognitive computing offers many features which justify its need in the banking sector.

1. **Enhancement of client service:** Cognitive computing systems analyze volumes of client data such as customer transaction history, spending patterns etc. Thus, customized recommendations can be provided for products such as saving accounts, credit cards, loan options etc. These recommendations are based on customers' financial condition, individual needs and risk profile. Therefore, overall proactive help, and personalization services are provided. This improves customer trust and satisfaction.
2. **Fraud detection:** Banks can quickly identify possible fraud by using cognitive computing to spot patterns and abnormalities in transactional data. The transactional data includes credit card transactions, online internet banking activities, ATM withdrawals, and payment through mobile phones. The cognitive computing systems utilize ML algorithms to recognize patterns and trends in transaction data. Through ML algorithms, it is possible to analyze spending behavior, number of times transactions happened, geographic locations and merchant categories Thus, fraudulent activities can be detected through the evaluation of unusual transaction behavior. Cognitive banking enables real-time monitoring of transactional data to detect fraudulent activities. Cognitive banking systems continuously learn through transactional behavior and detect fraud. Depending on the risk and severity levels of fraudulent activity, the alert system is optimized and prioritized. Thus, cognitive banking helps to mitigate potential losses and identify money laundering activities, terrorist financing etc.
3. **Risk management:** Cognitive computing helps to analyze transactional data including credit card reports, historical transactions to evaluate credit worthiness of the customers. By predicting, the system forecasts lending decisions, emerging risks, and allows proactive measures to mitigate losses. The system can provide early warning signs of potential risks to the bank. The system alerts the banks with investment strategies based on financial market fluctuations. The cognitive system helps the banks to meet compliance requirements given by government bodies.
4. **Improved decision-making:** By social media views, customer reviews, and feedback, cognitive systems identify emerging opportunities and threats for business. Bank personnel understand customer preferences, areas of improvement, cyber security threats and enhance customer satisfaction. Through news, social media posts and use of cognitive computing, bank personnel receive timely alerts about market disruptions, regulatory updates, emerging risks, which helps in making faster decisions.

5. **Automation of routine activities:** Cognitive systems automate routine tasks such as data input, document and customer inquiry processing. Cognitive systems automate the process of data entry by removing the need for traditional content collection, such as, paper documents, electronic records, and digital forms. Cognitive systems, through optical character recognition analysis and historical data patterns, improve data accuracy, and reduce manual errors. Cognitive computing uses NLP to categorize documents, then summarize and extract key information. This enables banks to analyze loan agreements and compliance forms easily.

   By automating routine tasks like data input, document processing, and customer questions, cognitive computing frees human workers to concentrate on higher-value work.
6. **Regulatory compliance:** By evaluating regulatory papers, keeping an eye on compliance-related activities, and identifying any infractions, cognitive computing systems can assist banks in navigating the complexities of regulatory requirements.
7. **Predictive analytics:** With the help of cognitive computing, banks can more precisely predict consumer behavior, industry trends, and financial performance. This supports investment strategies, marketing efforts, and product development.
8. **Financial guidance:** Through an analysis of individual financial information and inclinations, cognitive systems are able to provide tailored guidance on financial objectives such as retirement planning, investing, saving, and budgeting.

10.4 ROLE OF AI / ML FOR COGNITIVE BANKING

AI serves as the foundation of cognitive computing, providing the algorithms, models, and techniques necessary to emulate human-like intelligence in machines. Through NLP, ML, computer vision, knowledge representation, reasoning, and decision-making, AI empowers cognitive computing systems to understand, analyze, and respond to complex information and tasks, ultimately augmenting human capabilities and facilitating intelligent automation across diverse domains [9–11].

The integration of AI into cognitive computing within the banking industry has transformed how financial sector functions, delivery services, and interact with customers. Here is a detailed explanation of the contribution of AI in cognitive computing within the banking industry:

- **Customer service and interaction:**
  - *Virtual assistants:* With the use of cognitive computing, it handles customer queries such getting account information, details of transactions, and thus offer personalized recommendations.

These systems use NLP and reduce the burden on human customer service representatives.
  - *Voice assistants:* Voice-activated AI assistants, such as those integrated into mobile banking apps or smart speakers, enable customers to perform banking tasks through voice commands. AI technologies allow these assistants to interpret spoken requests and perform transactions securely.
- **Fraud detection and prevention:**
  - *Anomaly detection:* The cognitive algorithms examine vast amount of transaction data to identify unusual patterns or anomalies for fraudulent behavior. By monitoring suspicious behavior and promptly identifying potential fraudulent incidents, AI-powered systems help banks mitigate risks and protect customers' assets.
  - *Behavioral biometrics:* AI-driven behavioral biometrics systems analyze user interactions with digital banking platforms to establish unique behavior profiles for each customer. These systems detect deviations from normal behavior patterns, such as unusual login times or transaction locations, to identify potential fraud attempts and prevent unauthorized access.
- **Risk management:**
  - *Credit scoring and loan approval:* AI algorithms assess creditworthiness, predict loan default risks, and automate the loan approval process by analyzing various data sources, including financial history, credit scores, and transaction behavior. By leveraging ML techniques, banks can make decisions for lending, the loan application process, and offer personalized loan products to customers.
  - *Portfolio management:* AI-powered predictive analytics models analyze trends in the market and customer options to optimize investment portfolios and mitigate investment risks. These models provide insights into asset allocation, portfolio diversification, and investment strategies, helping banks maximize returns and minimize losses.
- **Personalized banking:**
  - *Customer segmentation and targeting:* Cognitive-driven analytics tools segment bank customers based on their demographic information, financial behavior, and transaction history. By grasping customer preferences and requirements, banks can tailor product offerings, marketing campaigns, and personalized recommendations to enhance customer engagement and satisfaction.
  - *Recommendation engines:* Offer relevant banking products and services, such as savings accounts, credit cards, or investment options to customers.

- **Automation of processes:**
  - *Robotic process automation (RPA):* AI-powered RPA systems automate banking operations, such as inputting, processing documents, and account reconciliation. By reducing human errors and streamlining workflows, RPA enhances operational efficiency, reduces costs, and accelerates time-to-market for new products and services.
  - *Compliance and regulatory reporting:* Cognitive technologies automate compliance checks, regulatory reporting, and audit processes by analyzing transaction data for compliance violations and identifying potential regulatory risks. These systems ensure that banks comply with regulatory standards, reduce risks, and prevent expensive penalties.
- **Trading:**
  - *Trading strategies:* Cognitive algorithms analyze market information, feeds, and social media trends to identify opportunities and optimize investment strategies in trade. Automated trading systems execute trades based on predefined rules and algorithms, leveraging the ability to rapidly process large volumes of data and make instantaneous decisions in dynamic market conditions.
  - *Risk management:* AI-driven risk management models assess market risks, liquidity risks, and credit risks associated with algorithmic trading activities. These models monitor trading positions, measure portfolio exposures, and implement risk mitigation strategies to protect banks from financial losses and regulatory scrutiny.
- **Regulatory compliance:**
  - *Know Your Customer (KYC) compliance:* Cognitive systems analyze customer data, transaction histories, and risk indicators to detect malicious activities. These systems identify unusual patterns, flag high-risk transactions, and generate alerts for further investigation by compliance officers.
  - *Privacy and security:* AI technologies enhance data privacy and security in banking operations by identifying potential security threats, detecting data breaches, and implementing advanced encryption and authentication mechanisms. The cognitive systems react to security incidents in real-time to secure crucial customer assets and protect against cyberattacks.

10.5 COMPONENTS OF COGNITIVE COMPUTING IN BANKING

In the banking sector, cognitive computing encompasses various components, technologies, and applications tailored to address industry-specific

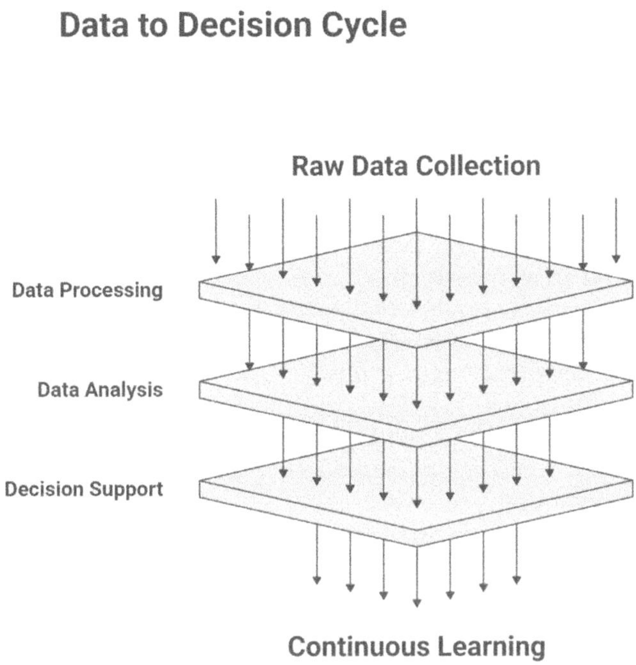

Figure 10.3 Components of Cognitive Banking.

challenges and opportunities [12–14]. Figure 10.3 shows components of cognitive banking.

- *Data acquisition:* Sources of data relevant to banking operations, including transaction records, customer information, market data, social media feeds, and regulatory documents.
- *Data processing and analysis:* Techniques and algorithms for processing and analyzing banking data, including NLP, ML, data mining, and analytics.
- *Decision support systems:* Tools and systems that assist bankers in making informed decisions by providing insights, recommendations, and predictions based on data analysis and modeling.
- **Role of technologies in cognitive systems for banking:**
  - *Natural language processing (NLP):* Technologies that interpret and process natural language data, such as chatbots, virtual assistants, sentiment analysis tools, and language translation services used in customer service, marketing, and communication.

  - *Machine learning (ML):* Algorithms and models that analyze banking data to detect patterns and predict customer behavior.
  - *Predictive analytics:* Techniques for forecasting future trends, identifying emerging risks, and anticipating customer needs based on historical data and predictive models.
  - *Computer vision:* Technologies for analyzing visual data in banking operations, including document recognition, signature verification, and facial recognition used in identity verification and security systems.
  - *Knowledge graphs and ontologies:* Graph-based representations of banking knowledge and relationships between entities, used in knowledge management, recommendation systems, and risk analysis.
  - *Robotic process automation (RPA):* Software bots that automate routine, rule-based and regulatory reporting in banking operations.
- **Applications of cognitive computing in banking:**
  - *Customer experience enhancement:* Applications that improve customer engagement, satisfaction, and loyalty through personalized banking experiences, proactive communication, and responsive customer service.
  - *Fraud detection and prevention*: Systems that analyze banking data to detect fraudulent activities, identify anomalies, and prevent financial losses through real-time monitoring, transaction analysis, and behavior profiling.
  - *Personalized banking services:* Tools and services that deliver tailored personalized products based on individual customer profiles, preferences, and financial objectives.
  - *Operational efficiency improvement:* Applications that streamline banking processes, optimize resource allocation, and reduce costs through automation, optimization, and workflow management.
  - *Algorithmic trading and investment:* Technologies that analyze market data, execute trades, and optimize investment portfolios based on quantitative models, algorithmic strategies, and risk management techniques.
  - *Regulatory reporting and governance:* Systems that ensure compliance with banking regulations, monitor for compliance violations, and facilitate regulatory reporting, audits, and examinations.
  - *Cybersecurity and data privacy:* Solutions that protect against cyber threats, safeguard sensitive data, and ensure data privacy and confidentiality through encryption, authentication, and security monitoring.

By categorizing cognitive computing components, technologies, and applications specific to the banking sector, this taxonomy provides a structured framework for understanding and implementing cognitive computing solutions tailored to banking industry needs.

10.6 CHALLENGES IN COGNITIVE BANKING

Data serves as a fundamental driver of technological advancements. However, the presence of erroneous data represents a significant concern, highlighting the critical importance of data accuracy. The scarcity of individuals possessing requisite data skills presents a formidable obstacle, exacerbated by the shortage of proficient data scientists. They endeavor to equip existing personnel with AI expertise is a time-intensive process, compounded by the challenge of recruiting skilled professionals in this burgeoning field [15].

Furthermore, while data privacy regulations play a pivotal role in overseeing AI applications, their current efficacy in regulating AI remains somewhat constrained. The onus of ensuring data access and privacy rests squarely with management, underscoring the paramount importance of stringent privacy measures.

The efficacy of applications such as 'speech to text' and 'text to speech' is contingent upon robust NLP capabilities. Consequently, the institution of proper protocols for data collection, validation, standardization, rectification, and dissemination is imperative.

The substantial costs associated with installation, maintenance, and upgrades, coupled with prolonged downtimes for error rectification, underscore the monumental task of managing vast datasets effectively. Additionally, the necessity of modeling techniques to derive precise model parameters from limited data samples cannot be overstated.

The pervasive use of AI can engender ethical, moral, and legal dilemmas, representing a significant challenge in its widespread adoption. Furthermore, the transformative nature of AI can potentially disrupt traditional skill sets, necessitating adaptation to dynamic process changes.

Despite these challenges, AI presents a myriad of opportunities, particularly in the banking sector, where it fosters automation and augments customer satisfaction. Regulatory bodies must prioritize data privacy safeguards to shield clients' sensitive information, while collaborative efforts with academic institutions can help cultivate a pipeline of skilled data scientists. Establishing dedicated AI roles within banking institutions can further accelerate the sector's growth trajectory.

10.7 INDIAN BANKING SECTOR AND COGNITIVE COMPUTING

The Indian banking sector leverages cognitive computing to enhance revenue, improve customer experience, optimize cost structures, and

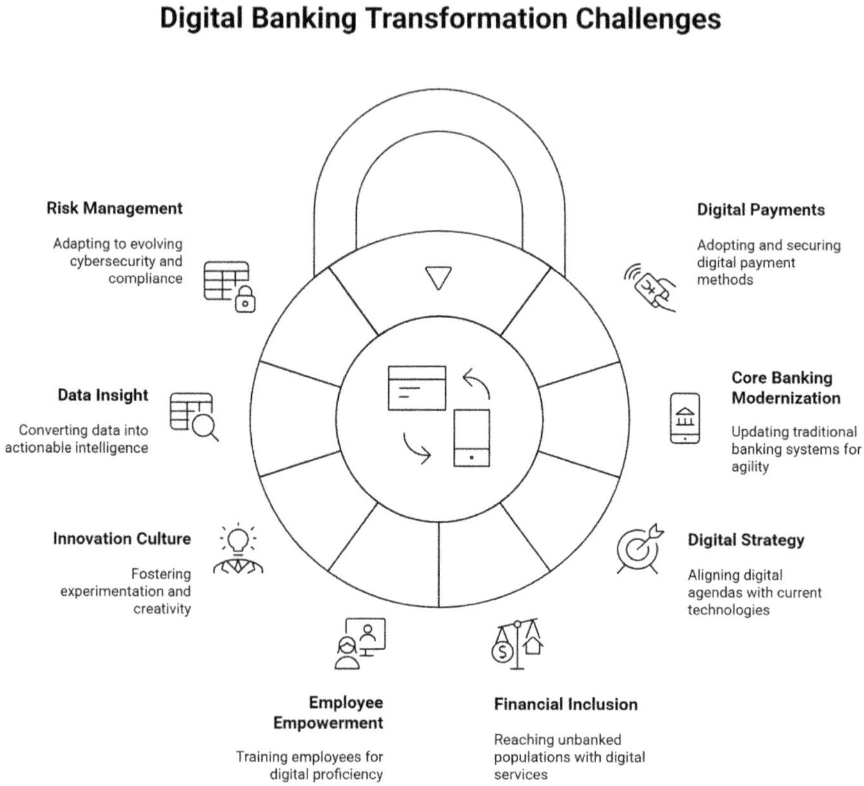

FIGURE 10.4 Challenges Faced in the Banking Sector.

manage enterprise risk. However, the operational capabilities of banks in implementing these solutions may vary, reflecting differences in their technology agendas [16–18].

The key challenges in the banking industry includes increasing core banking value, transitioning from information to insight, adapting to changing risk regimes, embracing electronic payment methods, addressing financial inclusion, empowering employees, and fostering innovation, as shown in Figure 10.4.

Salient features of technology adoption in Indian banking include:

- *Internet banking:* India has a rapidly growing, internet-savvy young consumer base.
    - *Business intelligence (BI):* The adoption of BI and analytics is aimed at increasing overall profitability in a competitive environment

- **Customer management:**
  - *Risk management and information security*: The use of cognitive computing mitigated the risks for credit systems and cyber security threats.
- **Technology in training and e-learning:**
  - *Financial inclusion:* Technology-enabled financial services promise to enable millions to access banking services, necessitating focus on banking and payment channels, technology platforms, and regulatory frameworks.
  - *Mobile banking:* Mobile banking has become a primary focus for banks, enhancing customer engagement in urban regions and advancing financial inclusion in rural areas.
  - *Payment systems:* India has experienced a transition from traditional cash payments to contemporary electronic payment systems. Private sector banks are at the forefront of electronic transactions, utilizing debit cards, credit cards, and internet banking.

10.8 CONCLUSION

The necessity for cognitive computing in banking arises from the need to handle vast amounts of data, provide better customer service, and remain competitive in a rapidly changing financial environment. Cognitive computing plays a significant role in improving the system of banking. AI and ML play pivotal roles for enhancing customer service, and decision-making. But it still faces many challenges that include data privacy concerns, integration with existing systems and addressing biases in AI algorithms.

In the Indian banking sector, cognitive computing enables a shift from traditional cash-based systems to modern electronic payment methods. Despite challenges, the future of cognitive banking looks promising, with continuous advancements expected to further integrate cognitive technologies into banking operations, ultimately reshaping the financial industry.

REFERENCES

[1] V. N. Gudivada, S. Pankanti, G. Seetharaman and Y. Zhang, "Cognitive Computing Systems: Their Potential and the Future", *Computer*, vol. 52, no. 5, 2019, pp. 13–18. https://doi.org/10.1109/MC.2019.2904940.

[2] C. R. Megha, A. Madhura and Y. S. Sneha, "Cognitive Computing and its Applications", *2017 International Conference on Energy, Communication, Data Analytics and Soft Computing (ICECDS)*, Chennai, India, 2017, pp. 1168–1172, https://doi.org/10.1109/ICECDS.2017.8389625.

[3] M. Chen, F. Herrera and K. Hwang, "Cognitive Computing: Architecture, Technologies and Intelligent Applications", *IEEE Access*, vol. 6, 2018, pp. 19774–19783. https://doi.org/10.1109/ACCESS.2018.2791469.

[4] J. O. Gutierrez-Garcia and E. López-Neri, "Cognitive Computing: A Brief Survey and Open Research Challenges", *2015 3rd International Conference on Applied Computing and Information Technology/2nd International Conference on Computational Science and Intelligence*, Okayama, Japan, 2015, pp. 328–333, https://doi.org/10.1109/ACIT-CSI.2015.64.

[5] J. L. C. Sanz and Y. Zhu, "Toward Scalable Artificial Intelligence in Finance", *2021 IEEE International Conference on Services Computing (SCC)*, Chicago, IL, USA, 2021, pp. 460–469, https://doi.org/10.1109/SCC53864.2021.00067.

[6] Spherical Insights LLP, "Global Cognitive Computing Market Size, Share and COVID-19 Impact Analysis, By Technology ... Analysis and Forecast 2021-2030", 2022, https://www.sphericalinsights.com/reports/cognitive-computing-market

[7] M. L. Fannisha, H. Hamzah, T. K. Ahsyar and E. Saputra, "Usability Evaluation of Flip Applications Using Cognitive Walkthrough and Think Aloud Evaluation Methods", *2023 2nd International Conference on Smart Technologies and Systems for Next Generation Computing (ICSTSN)*, Villupuram, India, 2023, pp. 1–6, https://doi.org/10.1109/ICSTSN57873.2023.10151607.

[8] Longbing Cao, "AI in Finance: Challenges, Techniques and Opportunities", *arxiv*, 2021.

[9] Harry, Thomas, James Thomas, and Nadira Hatim, "*Cognitive Banking: AI/ML, Data Integrity, and the Future of Financial Services*". Available at: https://www.researchgate.net/publication/373361180_Cognitive_Banking_AIML_Data_Integrity_and_the_Future_of_Financial_Services

[10] S. Biswas, B. Carson, V. Chung, S. Singh, and R. Thomas, "AI-Bank of the Future: Can Banks Meet the AI Challenge", *McKinsey & Company*, September 2020. Available at: https://www.mckinsey.com/industries/financial-services/our-insights/ai-bank-of-the-future-can-banks-meet-the-ai-challenge

[11] A. Aaliyah , D. Khan, and S. Poblete, *Innovating Banking with AI/ML: A Data Integrity and Governance Imperative*, 2023. Retrieved from https://www.researchgate.net/publication/373297136_Innovating_Banking_with_AIML_A_Data_Integrity_and_Governance_Imperative

[12] A. Agarwal, C. Singhal, and R. Thomas, "AI-Powered Decision Making for the Bank of the Future", *McKinsey & Company*, March 2021. https://www.mckinsey.com/industries/financial-services/our-insights/ai-powered-decision-making-for-the-bank-of-the-future

[13] S. R Guruvayur, "*Cognitive Banking Architecture – Human Centric AI framework for automated Customer Engagement in Banking*", Bankbuddy.ai.

[14] O. H. Fares, I. Butt, and S. H. M. Lee, "Utilization of Artificial Intelligence in the Banking Sector: A Systematic Literature Review", *Journal of Financial Services Marketing*, vol. 28, 2023, pp. 835–852. https://doi.org/10.1057/s41264-022-00176-7

[15] Onespan whitepaper, "Risk Analytics for Fraud Prevention: Top Use Cases in Banking", 2020, www.onespan.com/sites/default/files/2020-10/OneSpan-WhitePaper-A4-Risk-Analytics-Fraud-Prevention_20200930_1.pdf
[16] M. Nawaz1, Prof. Triveni. K2, Prof. Bharathi, and G. R2, "Application of Artificial Intelligence in Indian Banking-Opportunities and Challenges," *International Journal of Trend in Scientific Research and Development (IJTSRD)*, vol. 5, no. 1, November-December 2020, pp. 429–432.
[17] N.V. Kavitha, "Banktech Adoption Opportunities and Challenges in India – An Empirical Study", *IOSR Journal of Business and Management (IOSR-JBM)*, e-ISSN: 2278-487X, p-ISSN: 2319-7668. vol. 21, no. 3. Ser. III (March. 2019), pp. 50–54.
[18] Tata Consultancy Services, "*Computer Vision: A Game Changer for Banking, Financial Services and Insurance*", whitepaper www.tcs.com/content/dam/global-tcs/en/pdfs/insights/whitepapers/Computer-Vision-in-Banking.pdf

Chapter 11

Investigating the Role of Large Language Models in Cognitive Computing

Ujwala Bharambe, Sangita S. Chaudhari, Manimala Mahato, Kaushal Patil, and Rekha Ramesh

11.1 INTRODUCTION

Cognitive computing is an advanced field within artificial intelligence (AI) that focuses on creating systems capable of simulating human thought processes. These systems integrate various AI components, including machine learning (ML), natural language processing (NLP), and neural networks, to analyze complex, unstructured data and make decisions based on that analysis (Aghav-Palwe & Gunjal, 2021). Unlike traditional computing systems that operate on predefined rules, cognitive computing systems learn from interactions and continuously adapt their behaviors over time. This allows them to process and understand vast amounts of data in a way that mimics human cognition, making them particularly valuable for applications requiring real-time decision-making and personalized user interactions (Chen et al., 2018).

11.1.1 Cognitive Computing

Cognitive computing involves the ability to reason, learn, and interact naturally with humans. Reasoning enables cognitive systems to interpret data, make inferences, and provide logical solutions to problems. For example, a cognitive system can use data from a customer survey to identify patterns in customer behavior and suggest ways to improve customer experiences. Learning is achieved through ML algorithms that allow these systems to improve their performance-based on new data and experiences. For instance, a cognitive system can learn to recognize objects from a set of images by training the system on thousands of labeled images. Interaction with humans is achieved through the use of NLP and natural language understanding (NLU), which allow cognitive systems to understand and respond to natural language queries. NLP, a critical component of cognitive computing, allows these systems to understand, interpret, and generate human language. This facilitates more natural interactions between humans and machines.

 DOI: 10.1201/9781003501220-11

Table 11.1 Examples of Cognitive Systems

Parameter	*IBM Watson*	*Google Assistant*	*Microsoft Cortana*	*Amazon Alexa*	*IBM Watson for Oncology*
Natural Language Processing (NLP)	✓	✓	✓	✓	✓
Machine Learning Capabilities	✓	✓	✓	✓	✓
Personalization and Learning from User Interactions	✓	✓	✓	✓	✗
Voice Interaction	✗	✓	✓	✓	✗
Domain-Specific Expertise	✓	✗	✗	✗	✓

Chen et al. (2018) emphasize that cognitive computing systems are designed to handle real-world data, providing more comprehensive and contextually relevant responses. Unlike traditional non-cognitive systems that rely on predefined rules and structured data, cognitive systems can process unstructured data and learn from it to adapt and improve their responses over time. Some examples of real-world cognitive systems are compared in Table 11.1.

There is a growing number of cognitive systems leading the way, including IBM Watson, Google Assistant, Microsoft Cortana, Amazon Alexa, and IBM Watson for Oncology (Hurwitz, Kaufman, Bowles, Nugent, Kobielus & Kowolenko, 2015; Khalil, Jones, Babar, Jan, Zafar & Alhussain, 2019. Using advanced NLP and ML capabilities, IBM Watson excels at understanding and analyzing large amounts of unstructured data, making it indispensable in fields like healthcare and finance. Through NLP and ML, Google Assistant offers personalized, voice-based interactions, enhancing user experiences by learning from individual interactions. Similarly, Microsoft Cortana uses these technologies to offer personalized assistance, integrating seamlessly into the Windows ecosystem to manage tasks and answer queries. Amazon Alexa, another voice-activated assistant, utilizes NLP and ML to control smart home devices and provide information, continuously learning to improve its responses. On the specialized end, IBM Watson for Oncology demonstrates the pinnacle of cognitive computing in healthcare, employing NLP and ML to analyze medical literature and patient data, thus aiding oncologists with evidence-based treatment recommendations.

The rise of cognitive computing systems challenges traditional assumptions about user-system interactions, such as the direction of the user-artifact relationship and functional transparency (Schuetz & Venkatesh, 2020). These systems can adapt to situations, perceive their environments, and assist in complex problem-solving across various industries, including healthcare, finance, and education (Schuetz & Venkatesh, 2020; Aghav-Palwe & Gunjal, 2021). As these systems continue to evolve, they offer opportunities for novel theory development in information systems research and have the potential to transform businesses and improve decision-making processes (Hurwitz et al., 2015; Schuetz & Venkatesh, 2020).

11.1.2 Large Language Models

Language stands as humanity's most potent tool, enabling the expression of thoughts and emotions and facilitating communication with others Pinker, 2003; Turing, 2009. However, machines lack the intrinsic ability to comprehend and communicate in human language, necessitating powerful AI algorithms to acquire such capabilities. The ultimate aim of AI research is the creation of machines capable of reading, writing, and conversing like humans Dwivedi, Kshetri, Hughes, Slade, Jeyaraj, Kar, ... & Wright, 2023; Zhao, Zhou, Tang, Wang, Hou, ... & Wen, 2023; Jin, Liu, Han, Jiang, Ji & Han, 2024. NLP emerges as the AI branch dedicated to realizing this goal.

At the core of NLP are language models, which predict or generate the probability or likelihood of linguistic units (e.g., words, phrases, sentences) based on context. The evolution of language models has progressed from initial statistical language models (SLMs) to neural language models (NLMs), then to pre-trained language models (PLMs), and finally to the current state of large language models (LLMs). Specifically, SLMs use simple probability distributions to model word sequences, while NLMs employ neural networks to grasp complex patterns and representations of language. PLMs leverage large-scale corpora and self-supervised learning to capture general linguistic knowledge. LLMs extend PLMs by incorporating massive data, computation, and advanced algorithms, resulting in more expressive, sophisticated, and adaptable language models (Hupkes, Giulianelli, Dankers, Artetxe, Elazar, Pimentel, ... & Jin, 2022; Wei, Wang, Schuurmans, Bosma, Chi, Le & Zhou, 2022).

Despite being trained on seemingly simple tasks like predicting the next word in a sentence, LLMs can capture much of the structure and meaning of human language. They also possess a remarkable amount of general knowledge about the world and can memorize an enormous number of facts during training and can be thought of as giant, flexible brains that can be taught to do almost anything, provided they have access to enough data and processing power. When you interact with ChatGPT, you are engaging with one of the most advanced pieces of AI technology in existence.

In recent years, significant advancements in NLP have been achieved with LLMs, such as the generative pre-trained transformer GPT series of models (Radford, Narasimhan, Salimans & Sutskever, 2018; Radford, Wu, Child, Luan, Amodei & Ilya Sutskever, 2019; Brown, Mann, Ryder, Subbiah, Kaplan, Dhariwal, Neelakantan et al., 2020; Achiam, Adler, Agarwal, Ahmad, Akkaya, Aleman, Almeida et al., 2023). Extensively trained on textual data, these models can generate human-level texts and perform language-based tasks with exceptional precision. Despite these advancements having greatly benefited various aspects of work and life, LLMs might not be widely understood by practitioners, particularly those without a background in NLP.

11.1.2.1 Evolution of Large Language Models

Large language models (LLMs) span decades, marked by significant milestones and technological advancements. Figure 11.1 shows a timeline showing significant milestones in the development of LLMs. Starting with early experiments in the 1950s, researchers at IBM and Georgetown University developed a system to automatically translate phrases from Russian to English, igniting a surge of interest in NLP and showcasing the potential for computers to process human language. This momentum continued into the 1960s with the introduction of Eliza by MIT researcher Joseph Weizenbaum, the world's first chatbot (Weizenbaum, 1966). Eliza used pattern matching and substitution methodology to simulate conversation, marking the beginning of NLP research and laying the groundwork for the development of more complex LLMs.

The 1990s saw the emergence of SLMs, which used probabilistic methods to model sentence likelihoods based on word sequences. This helped to capture syntactic structures and handle more complex corpora (Rosenfeld & Clarkson, 1997). In 1997, the introduction of Long Short-Term Memory (LSTM) networks by Hochreiter and Schmidhuber addressed the vanishing gradient problem in traditional recurrent neural networks (RNNs), enabling the creation of deeper neural networks capable of handling vast amounts of data (Hochreiter, 1997). The 2000s brought NLMs, which leveraged neural networks to predict word probabilities in sequences, utilizing word vectors to represent words in continuous vector space and improving context understanding (Wang, 1997). Key milestones in the field include the introduction of Word2Vec by Mikolov, Sutskever, Chen, Corrado and Dean in 2013, and Stanford's CoreNLP suite in 2010. These developments facilitated a range of NLP tasks, significantly advancing the field.

The introduction of transformer models by Vaswani, Shazeer, Parmar, Uszkoreit, Jones, Gomez, ... and Polosukhin. in 2017 revolutionized NLP by relying on self-attention mechanisms to train efficiently on large datasets. This innovation enabled models like Generative Pre-trained Transformer

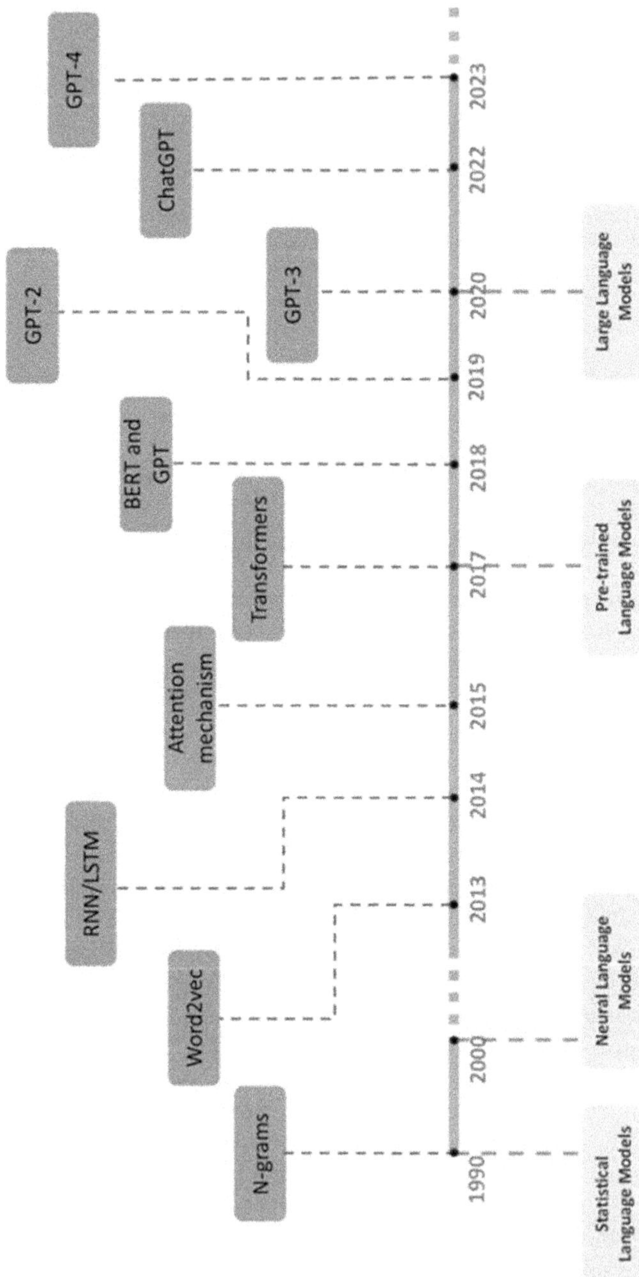

Figure 11.1 Timeline Showing Significant Milestones in the Development of LLMs.

(GPT) and Bidirectional Encoder Representations from Transformers (BERT) to generate and classify text effectively. From 2018 onwards, the focus shifted to building larger models, such as BERT and GPT-2, which demonstrated significant breakthroughs in understanding context and generating coherent text. Pre-trained Language Models (PLMs) and LLMs, such as GPT-3 and GPT-4, further pushed the boundaries of NLP, consolidating pre-training on vast corpora with fine-tuning on specific tasks, resulting in adaptability and performance across various domains (Chen, 1999).

Recent advancements in LLMs include powerful models such as GPT-3, Text-To-Text Transfer Transformer (T5), and GPT-4, which are compared in the Table 11.2. The GPT-3, T5, and GPT-4 LLMS offer high accuracy and versatility, but require substantial computational resources. The Pathways Language Model (PaLM) and Generalist Language Model (GLaM) from Google improved scalability and efficiency, but they were complex to

Table 11.2 Comparison of Large Language Models

Model	*Year*	*Features*	*Capabilities*	*Limitations*
BERT	2018	Bidirectional encoder, pre-training on vast data	Contextual understanding, adaptable to various tasks	Computationally intensive, large model size
GPT-2	2019	Transformer architecture, extensive pre-training	Generate coherent and contextually relevant text	Can produce biased or harmful outputs without fine-tuning
T5	2020	Text-to-Text Transfer Transformer	Unified framework for NLP tasks, high versatility	High computational costs, requires fine-tuning
PaLM	2022	Pathways Language Model by Google	540 billion parameters, enhanced scalability	High computational demand, complex infrastructure
GPT-4	2023	Multimodal capabilities, extensive pre-training	Handles text and image inputs, superior performance	Requires massive computational power, expensive
LaMDA	2023	Language Model for Dialogue Applications by Google	Focused on conversation. quality and coherence	Context management challenges, requires extensive data
Claude	2024	Large language model by Anthropic	Emphasis on alignment with human values and ethics	Still emerging, ongoing refinement needed
GPT-5	2024	Advanced multimodal, enhanced capabilities	Further improved text, image, and possibly video processing	Extreme computational requirements, high costs

deploy and train. The latest models, such as Claude and GPT-5, emphasize ethical alignment and advanced multimodal capabilities, but they are still challenged by high computational demands.

11.2 LANDSCAPE OF COGNITIVE COMPUTING

In cognitive computing, self-learning systems use a wide range of technologies, such as the Internet of Things (IoT), big data analysis, cloud computing, data mining, pattern recognition, NLP, computer vision deep learning (DL), reinforcement learning, etc., to mimic the way the brain works. These cognitive systems are designed to handle complex problems that traditionally require human intervention by processing vast amounts of structured and unstructured data through ML/DL algorithms. These systems analyze data, they continually improve their ability to recognize patterns, understand context, and generate insights(Dhanalakshmi, & Sri Devi, 2020). Over time, cognitive computing systems develop the capability to anticipate new problems and propose potential solutions, making them invaluable in fields such as healthcare, finance, and customer services. For instance, in the healthcare industry, cognitive computing systems can analyze patient data to detect potential issues, suggest treatments, and predict patient outcomes (Sabarmathi & Leelavathi, 2019).

Cognitive computing systems are (1) adaptive, (2) interactive, (3) iterative and stateful, and (4) contextual in nature. Adaptive systems can learn from data and adjust their algorithms based on new information. This adaptability ensures that cognitive computing systems remain relevant and accurate even as the data landscape changes. For example, in healthcare, cognitive systems can adapt to evolving medical research and patient data, improving diagnostic accuracy and treatment plans over time. Interactive systems can understand and respond to human input using technologies like NLP, enabling seamless communication between humans and machines. This interactivity is crucial for applications like virtual assistants and customer service bots, where understanding and responding to user queries accurately is essential. For instance, a healthcare chatbot might use NLP to understand a user input such as "I have a sore throat" and suggest treatments such as "take an over-the-counter pain reliever" or "see your doctor for a check-up".

Iterative and stateful features allow cognitive systems to maintain feedback. They also retain information about past interactions to inform future ones. This iterative learning process is similar to human learning, where experiences shape future decisions and behaviors. In a customer service context, a cognitive system can learn from previous interactions to provide more accurate and personalized responses in future engagements. Contextual capabilities enable these systems to understand the context in which data is being used, ensuring that the insights generated are relevant and applicable. For example, a cognitive system can be trained to recognize a customer's voice and provide tailored responses to them, based on their past interactions with the company. By integrating these features, cognitive computing systems provide a powerful tool for solving complex problems and enhancing decision-making across various domains.

11.2.1 Key Advancement and Technologies in the Field of Cognitive Computing

In cognitive computing, self-learning systems use a wide range of technologies, such as IoT, big data analysis, cloud computing, data mining, pattern recognition, NLP, computer vision, deep learning (DL), reinforcement learning, etc., to mimic the way the brain works. Figure 11.2 illustrates key technologies in the field of cognitive computing. These cognitive systems are designed to handle complex problems that traditionally require human intervention. They process vast amounts of structured and unstructured data through ML/DL algorithms. As these systems analyze data, they continually improve their ability to recognize patterns, understand context, and generate insights (Dhanalakshmi & Sri Devi, 2020). Over time, cognitive computing systems develop the capability to anticipate new problems and propose potential solutions, making them invaluable in fields such as healthcare, finance, and customer services. For instance, in the healthcare industry, cognitive computing systems can analyze patient data to detect potential issues, suggest treatments, and predict patient outcomes (Sabarmathi & Leelavathi, 2019).

The progress and capabilities of cognitive computing domain have been significantly driven by the key advancements and technologies in this field. NLP innovations have led to improved algorithms and models that allow machines to understand and generate human language more effectively. ML and deep learning techniques, especially neural networks, have greatly enhanced cognitive computing capabilities, enabling more sophisticated data processing and decision-making.

The development of large language models (LLMs) such as GPT and BERT has revolutionized language processing and generation, allowing for more natural and context-aware interactions. Big data analytics plays a crucial role by enabling the real-time processing and analysis of large data volumes, providing insights that drive cognitive computing applications.

In the era of visual information, computer vision technologies have enabled machines to interpret and analyze images, leading to advancements in image recognition and analysis. Robotic Process Automation (RPA) has streamlined workflows by automating repetitive tasks, making cognitive systems more efficient.

Speech recognition and synthesis technologies have improved voice-activated systems, making them more accurate and user-friendly. Advances

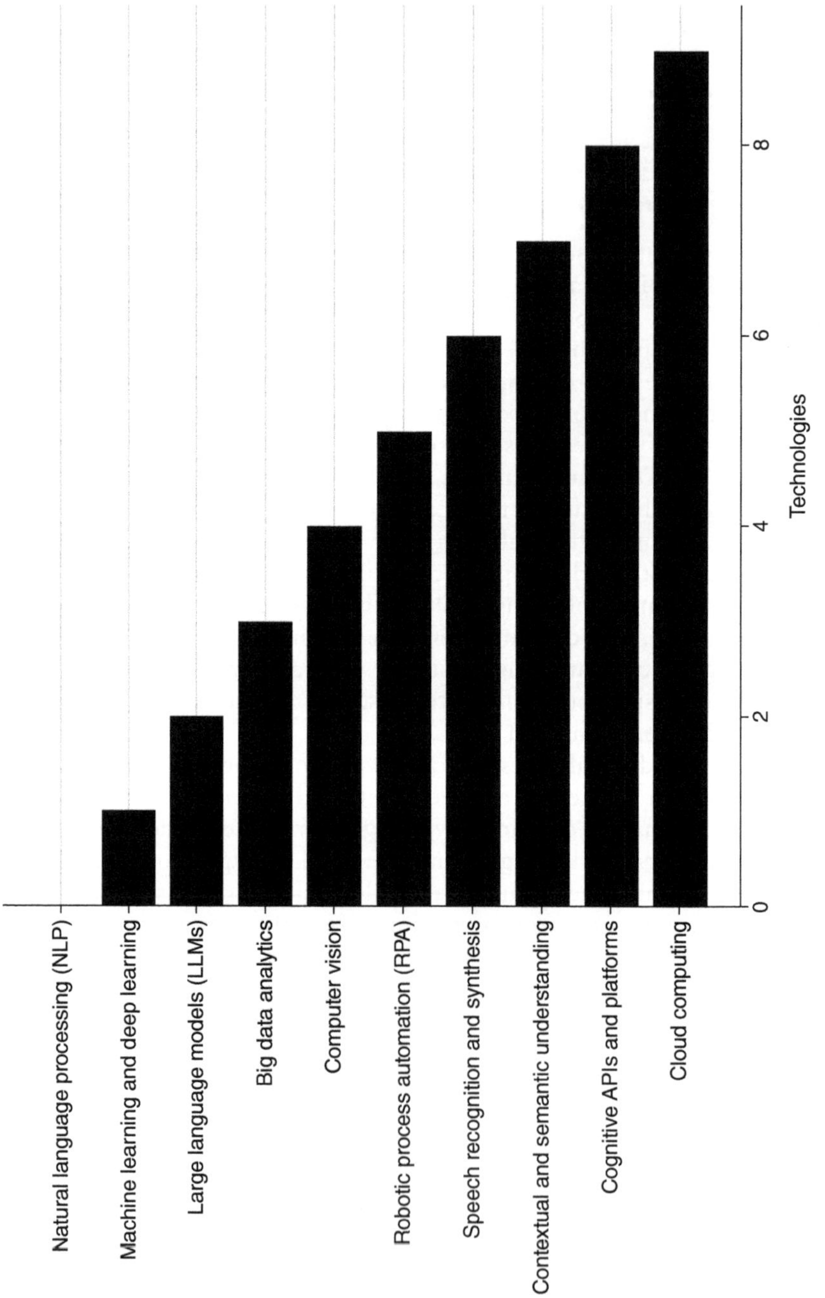

Figure 11.2 Key Technologies in the Field of Cognitive Computing.

in contextual and semantic understanding have enhanced the ability of cognitive systems to provide relevant and meaningful responses by better grasping the nuances of data.

The availability of cognitive APIs and platforms has simplified the integration of cognitive computing capabilities into various applications, making these advanced technologies more accessible to developers. Finally, cloud computing has provided the scalability and flexibility needed to deploy and manage cognitive computing solutions on a larger scale, further accelerating the growth and impact of this field.

11.3 OUTLINE OF LARGE LANGUAGE MODEL

LLM is a neural network with billions of parameters trained on vast amounts of unlabeled text using self-supervised or semi-supervised learning. These general-purpose models can perform a wide variety of tasks, from sentiment analysis to mathematical reasoning.

11.3.1 Architecture of a Typical Large Language Model

LLMs consist of multiple integral components that collectively empower them to comprehend, generate, and manipulate human language with exceptional fluency and precision. Figure 11.3 shows the architecture of a typical LLM. The embedding layer in a LLM is a critical component that transforms input tokens (such as words or subwords) into dense vector representations. These vectors serve as the initial input to the model, allowing it to work with numerical data that captures the syntactic and semantic properties of the tokens.

The input text is first tokenized into smaller units, such as words, sub words, or characters. Each token is assigned a unique identifier from a predefined vocabulary. The size of the vocabulary can vary, typically ranging from tens of thousands to millions of tokens. The core of the embedding layer is an embedding matrix, which is a learnable parameter of the model. The core of the embedding layer is an embedding matrix, which is a learnable parameter of the model. For each input token, the model performs a lookup in the embedding matrix to retrieve the corresponding dense vector. Each token is mapped to a dense vector of fixed size (the embedding dimension). These vectors capture semantic meanings, so that similar tokens have similar vector representations in the embedding space. The values in the embedding matrix are learned during training. As the model trains on large datasets, the embedding vectors adjust to reflect the semantic properties and relationships between tokens.

Since the transformer architecture does not inherently recognize the order of tokens, positional encoding is added to the embedding vectors to inject information about the position of tokens within a sequence. A unique

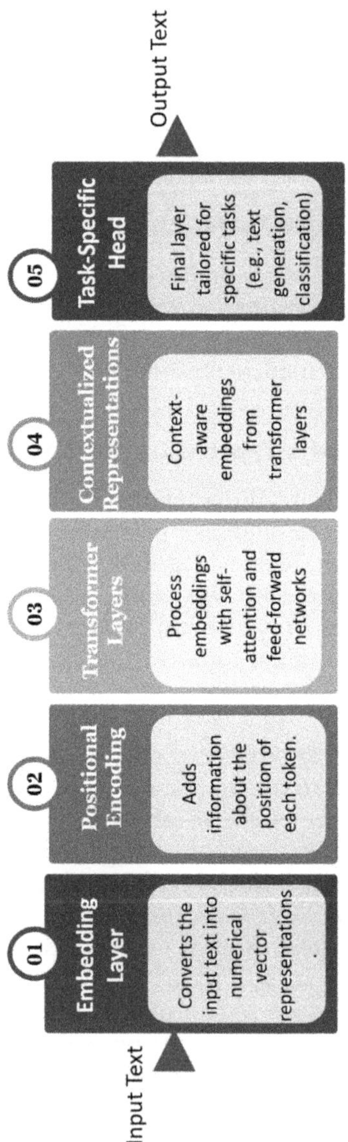

Figure 11.3 Architecture of a Typical Large Language Model.

positional encoding vector is added to each token's embedding vector, allowing the model to distinguish between tokens at different positions. This is crucial for understanding the structure and meaning of sentences.

The output of the embedding layer is a sequence of dense vectors, each representing a token in the input sequence. This sequence of embedding vectors is then fed into the subsequent layers of the transformer (such as self-attention layers) for further processing.

The transformer layer is the core component of LLM. The architecture is built on multiple stacked transformer layers, each consisting of two main sub-layers: Multi-Head Self-Attention Mechanism and Feed forward Neural Network (FFN). The self-attention mechanism allows the model to weigh the importance of different words in a sentence relative to each other. It works by taking the input sequence and computing a set of attention scores for each word in the context of all other words. This allows the model to focus on different parts of the sentence when processing each word, capturing relationships regardless of their distance in the sequence. Instead of calculating a single attention score for each word pair, the multi-head mechanism computes multiple sets of attention scores in parallel. The outputs of these multiple heads are then concatenated and linearly transformed to produce the final output. After the self-attention mechanism, the output is passed through a fully connected feed forward network. This layer is applied independently to each position, adding a non-linear transformation to the model. Each sub-layer is followed by layer normalization and a residual connection. The residual connection adds the input of the sub-layer to its output, helping to prevent vanishing gradients and enabling better flow of gradients during back propagation. Layer normalization helps in stabilizing and speeding up the training process.

After passing through all the stacked transformer layers, the model produces a final contextualized representation for each word in the sequence. This helps the model in handling polysemy, disambiguation and rich semantic understanding. This final representation is context-dependent and is used for various applications such as text generation, machine translation, sentiment analysis, question answering, etc.

The task-specific heads are additional layers or modules added on top of the core transformer layers to adapt the model for specific tasks. These heads are crucial when fine-tuning a pre-trained language model for a particular application. The task-specific head operates on top of the shared representations generated by the transformer layers. These layers provide a rich, contextually-aware understanding of the input, which the task-specific head refines and adapts for the specific task. The architecture allows different task-specific heads to be attached to the same pre-trained model. This flexibility enables the same underlying model to be fine-tuned and reused for a variety of tasks, making LLMs highly versatile.

11.3.2 Natural Language understanding by Large Language Models

Natural Language Understanding (NLU) is a crucial component of NLP that enables machines to comprehend, interpret, and generate human language. LLMs (which, to remind the reader, stands for large language models) like BERT, GPT-3, and T5 have revolutionized NLU by leveraging deep learning techniques, particularly transformer architectures. This detailed explanation covers the technical aspects of how NLU is performed by LLMs using a transformer. There are 5 important key concepts related to it; (1) transformer architecture, (2) pre-training techniques, (3) fine-tuning for specific tasks, (4) handling long texts

- **Transformer architecture:** The foundation of LLMs lies in the transformer architecture, which revolutionized NLP with its self-attention mechanism and encoder-decoder framework. The self-attention mechanism allows the model to weigh the importance of each word in a sentence relative to every other word, capturing long-range dependencies and contextual relationships (Vaswani et al., 2017). This architecture facilitates a deep understanding of language by enabling models to process and generate text more effectively than previous methods, such as recurrent neural networks (RNNs). The encoder-decoder framework, central to transformers, processes input text through the encoder and generates output through the decoder, making it versatile for various NLP tasks.
- **Pre-training and fine-tuning:** LLMs undergo a two-stage process: pre-training and fine-tuning. During pre-training, models like BERT and GPT-3 are trained on vast amounts of text data using techniques like masked language modeling and auto-regressive language modeling. MLM involves masking portions of the input text and training the model to predict the missing words, fostering a bidirectional understanding of context (Devlin, Chang, Lee & Toutanova, 2019). In the fine-tuning stage, these pre-trained models are adapted to specific tasks such as sentiment analysis or question answering by training them on smaller, task-specific datasets. This process ensures that the models achieve high performance across a variety of NLP applications by leveraging their extensive pre-trained knowledge.
- **Contextual embeddings:** Unlike traditional word embeddings, which provide static representations of words, LLMs generate contextual embeddings that capture the meaning of a word based on its context within a sentence. This approach addresses the issue of polysemy, where a single word can have multiple meanings. Contextualized representations allow the model to distinguish between different senses of a word, enhancing its ability to understand nuanced language

(Peters, Neumann, Iyyer, Gardner, Clark, Lee & Zettlemoyer, 2018). For example, the word "bank" will have different embeddings in the contexts of "river bank" and "financial bank," depending on the surrounding words.

- **Handling long texts:** One of the advancements brought by LLMs is their ability to handle long texts efficiently. Traditional transformers have a limitation on the input sequence length due to computational constraints. However, newer models like Longformer and Big Bird have introduced sparse attention mechanisms that extend the capability of transformers to process longer sequences without performance degradation (Beltagy, Peters, & Cohan, 2020; Zaheer, Guruganesh, Dubey Ainslie, Alberti, Ontanon, ... & Ahmed, 2020). These innovations are crucial for applications like document summarization and long-form text generation, where understanding and processing large amounts of text is necessary.
- **Task-specific applications:** LLMs are highly versatile and can be fine-tuned for a wide range of task-specific applications. For instance, in question answering tasks, models like BERT have achieved state-of-the-art performance by leveraging their deep understanding of bidirectional context (Devlin et al., 2019). Similarly, models like T5 excel in tasks such as text summarization and translation by converting all NLP tasks into a text-to-text format, making them incredibly flexible and powerful (Raffel, Shazeer, Roberts, Lee, Narang, Matena, ... & Liu, 2020). These applications demonstrate the broad utility of LLMs in real-world scenarios, enhancing various aspects of human-computer interaction.

11.4 INTERSECTION OF LLM AND COGNITIVE COMPUTING

This chapter discusses the evolution and architecture of cognitive computing, emphasizing the integration of various technologies such as LLMs. LLMs enhance cognitive computing systems by improving their ability to understand and generate human language. These models process natural language inputs, generate coherent responses, and translate languages, making cognitive systems more versatile and powerful. The paper highlights how LLMs contribute to the development of human-centered smart systems, including smart healthcare, affective interaction, and autonomous driving (Chen et al., 2018).

The intersection of LLMs (which, as a reminder to reads, stands for Large Language Models) and cognitive computing as shown in Figure 11.4 represents a convergence of advanced AI technologies designed to mimic, enhance, and augment human cognitive processes.

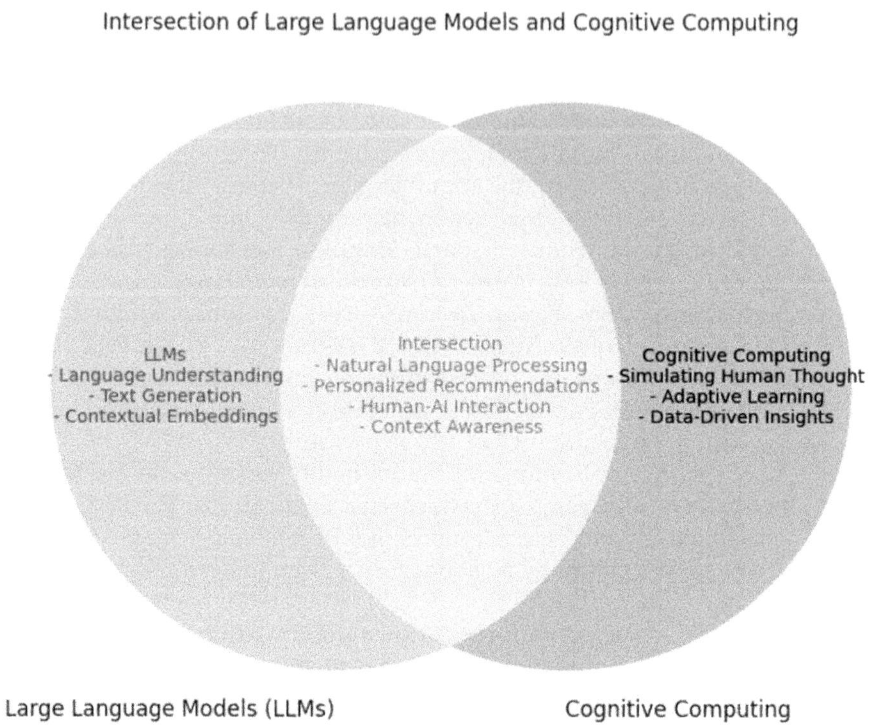

Figure 11.4 Intersection of Large Language Models and Cognitive Computing.

- Cognitive computing systems often rely on LLMs to process and understand vast amounts of natural language data, enabling more human-like interactions and decision-making. For instance, in customer service, LLMs can help cognitive systems understand and respond to customer inquiries in real-time.
- Combining LLMs with cognitive computing allows for systems that not only perform tasks with pre-trained knowledge but also adapt to new information dynamically. For example, an LLM embedded in a cognitive system could learn from user interactions to improve its responses over time.
- LLMs can enhance cognitive systems by providing contextually relevant information or generating hypotheses, while cognitive computing systems can provide the reasoning framework for decision-making, leading to more intelligent and context-aware AI applications.
- By integrating LLMs into cognitive computing platforms, the resulting systems can provide highly personalized and adaptive user experiences. For instance, in healthcare, such a system could tailor

advice and interactions based on a patient's unique medical history and preferences.

- LLMs face challenges around bias, interpretability, and ethical use, particularly in understanding and generating language in a socially responsible manner. Cognitive Computing also grapples with ethical issues, especially in ensuring that automated decision-making aligns with human values and is transparent. Addressing these challenges requires the combined strengths of both fields. LLMs can benefit from the ethical frameworks developed in cognitive computing, while cognitive systems can leverage the linguistic and contextual capabilities of LLMs to better understand the implications of their decisions.

The intersection of LLMs and cognitive computing offers promising avenues for creating more intelligent, adaptive, and human-like AI systems. By combining the language proficiency of LLMs with the learning, reasoning, and decision-making capabilities of cognitive computing, we can develop systems that not only understand and generate human language but also think, learn, and adapt like humans, opening up new possibilities in fields like healthcare, finance, customer service, and beyond.

11.5 CASE STUDY: INTEGRATION OF LLM-BASED COGNITIVE COMPUTING FOR CHEMO BRAIN REHABILITATION

The integration of cognitive computing and large language model (or, LLM) technology offers a transformative approach to addressing the cognitive impairments experienced by patients like Sarika, a breast cancer survivor undergoing chemotherapy. Chemo brain, characterized by memory lapses, difficulties in concentration, and reduced executive function, can severely impact daily life. To mitigate these effects, an advanced rehabilitation tool has been developed, leveraging the capabilities of cognitive computing and LLMs to provide a personalized and adaptive cognitive rehabilitation experience.

At the core of this tool is the use of LLMs, such as BERT and GPT-3, which excel in understanding and generating natural language (Devlin et al., 2019). These models facilitate seamless and intuitive interactions between the patient and the system. By incorporating spatio-temporal data, the tool tailors cognitive exercises to the patient's daily routines and specific environments. For example, in the morning when Sarika is more alert, the system can propose more challenging cognitive tasks. By contrast, during the afternoon, it might suggest combining physical activities with cognitive exercises in familiar environments like a park, thereby enhancing both mental and physical well-being (Vaswani et al., 2017). This contextual adaptation

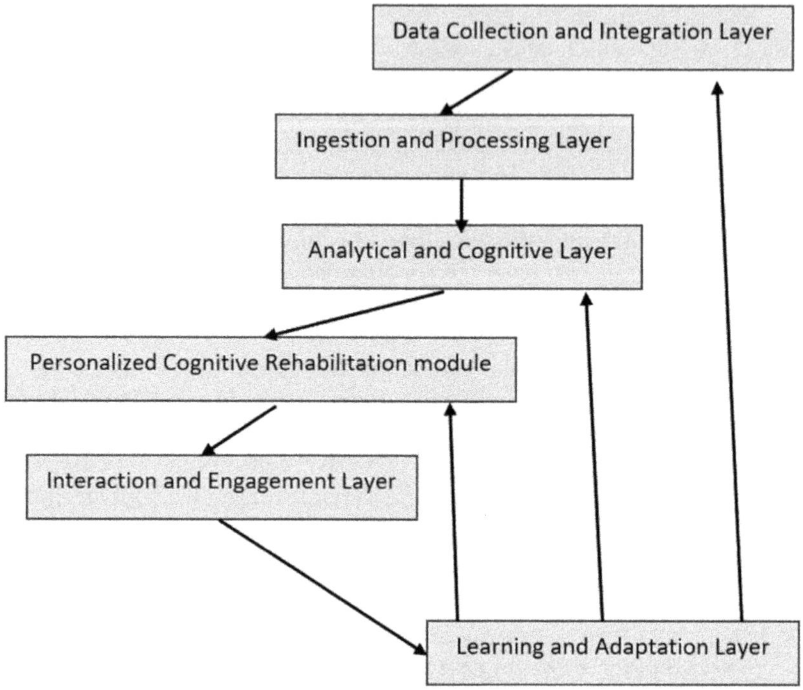

Figure 11.5 Proposed Architecture for LLM-based Cognitive Computing for Chemo Brain Rehabilitation Tool.

ensures that the rehabilitation program is both relevant and engaging, significantly improving its effectiveness (Figure 11.5).

The system also features a virtual health assistant powered by LLMs, enabling natural language interactions that are crucial for maintaining patient engagement and providing timely support. This assistant offers real-time reminders for medication, guides the patient through tailored cognitive exercises, and provides emotional support by responding to health-related queries and offering encouragement (Shneiderman, 2010). Furthermore, the cognitive computing framework enables continuous learning from the patient's interactions and performance data, dynamically adjusting the rehabilitation plan to ensure ongoing relevance and effectiveness (Goodfellow, Bengio, Courville & Yoshua, 2016). This integration of cognitive computing and LLM technology represents a sophisticated, adaptive, and personalized approach to managing chemo brain, offering significant improvements in the quality of life for patients undergoing chemotherapy.

The proposed rehabilitation tool for chemo brain patients integrates six core components in a cohesive and adaptive architecture. Starting with the

data collection and integration layer, it gathers patient data, spatio-temporal information, and health metrics. This data is then processed in the **ingestion and processing layer** to ensure it is clean and ready for analysis. The **analytical and cognitive layer** employs LLMs and cognitive computing algorithms to generate insights and personalized recommendations. These insights feed into the **personalized cognitive rehabilitation module**, which develops adaptive cognitive exercises tailored to the patient's needs. The **interaction and engagement layer** facilitates natural language interactions, making the system user-friendly and engaging. Finally, the **learning and adaptation layer** captures feedback and continuously refines the system, ensuring that it evolves based on patient performance and new data. Feedback loops between the learning and adaptation layer and other components ensure the system remains dynamic and responsive, providing personalized and effective support for cognitive rehabilitation.

Let us consider cancer patients Sarika, a breast cancer patient undergoing chemotherapy, has been struggling with cognitive impairments known as "chemo brain," which affect her memory, attention, and executive function, severely impacting her daily life. To help her, a new rehabilitation tool has been developed that integrates cognitive computing and LLMs. This tool offers personalized, adaptive, and context-aware cognitive exercises and support, using advanced technology to enhance Sarika's cognitive rehabilitation process.

The tool can employ LLMs such as BERT and GPT-5 to understand and generate human language, facilitating seamless interactions between Sarika and the system (Devlin et al., 2019). By incorporating spatio-temporal data, it customizes cognitive exercises to fit Sarika's daily routines and environments. For example, in the mornings when Sarika is more alert, the system suggests more challenging cognitive tasks. In the afternoons, it combines physical activities with cognitive exercises in familiar settings like a park, boosting both her physical and mental well-being. This high level of personalization ensures that the rehabilitation exercises are relevant, engaging, and effective in addressing Sarika's cognitive challenges (Vaswani et al., 2017).

11.6 CHALLENGES AND LIMITATIONS OF LLM-BASED COGNITIVE COMPUTING

Cognitive computing, which aims to simulate human thought processes in computerized models, faces a variety of challenges, especially when leveraging LLMs. One of the primary challenges is data quality and availability; cognitive systems depend on large datasets that must be accurate, relevant, and comprehensive. Incomplete or poor-quality data can lead to incorrect conclusions, and integrating diverse data sources into a cohesive dataset is complex. Performance, in this context, is critical; the accuracy of the solution, including the precision of intermediate stages, is paramount for effective cognitive computing. Furthermore, the complexity and scalability of these systems present significant hurdles. Developing cognitive systems capable of performing intricate tasks, understanding context, and adapting to new information is arduous. Ensuring these systems can handle large volumes of data and interactions without performance degradation is crucial for practical deployment. Efficiency, relating to the number of LLM inference calls required to solve a single problem, is vital due to the high computational cost associated with LLM inference. Another critical issue is the interpretability and transparency of cognitive systems. Many operate as "black boxes" with decision-making processes that are not easily understandable, which complicates trust and accountability. Additionally, ethical and bias concerns are prominent, as these systems can inherit biases from their training data, leading to biased outcomes in analyses and decisions. Flexibility in thought topology, allowing diverse structures like chains, trees, or graphs, enhances the capacity for divergent and creative thinking, which is particularly advantageous in addressing complex problems with multiple potential solutions. Addressing these challenges involves a combination of technological advancements, regulatory frameworks, ethical considerations, and interdisciplinary collaboration.

Despite these challenges, the benefits of LLM-based cognitive computing are substantial. One significant advantage is the enhancement of human-AI interaction, making it more intuitive and effective. By using advanced LLMs like BERT and GPT-3, cognitive systems can understand and generate human language, facilitating seamless and natural interactions. This is particularly beneficial in fields like healthcare, where LLMs can assist in diagnosing diseases, personalizing treatment plans, and providing emotional support to patients. The performance aspect ensures these interactions are precise and reliable. Another advantage is the ability of these systems to continuously learn and adapt from new data, improving over time and becoming more accurate and efficient, minimizing the number of LLM inference calls needed. This adaptability is crucial in dynamic fields where new information continuously emerges. Furthermore, cognitive computing systems can process and analyze vast amounts of data far more quickly and accurately than humans, providing valuable insights that can drive innovation and efficiency. For example, in finance, these systems can detect fraudulent activities in real-time, enhancing security and trust. Flexibility in organizing thoughts and responses allows these systems to address complex, multifaceted problems effectively. However, to fully realize these benefits, it is essential to address the underlying challenges, ensuring that cognitive computing systems are reliable, ethical, and transparent, fostering greater trust and widespread adoption in various industries(Figure 11.6).

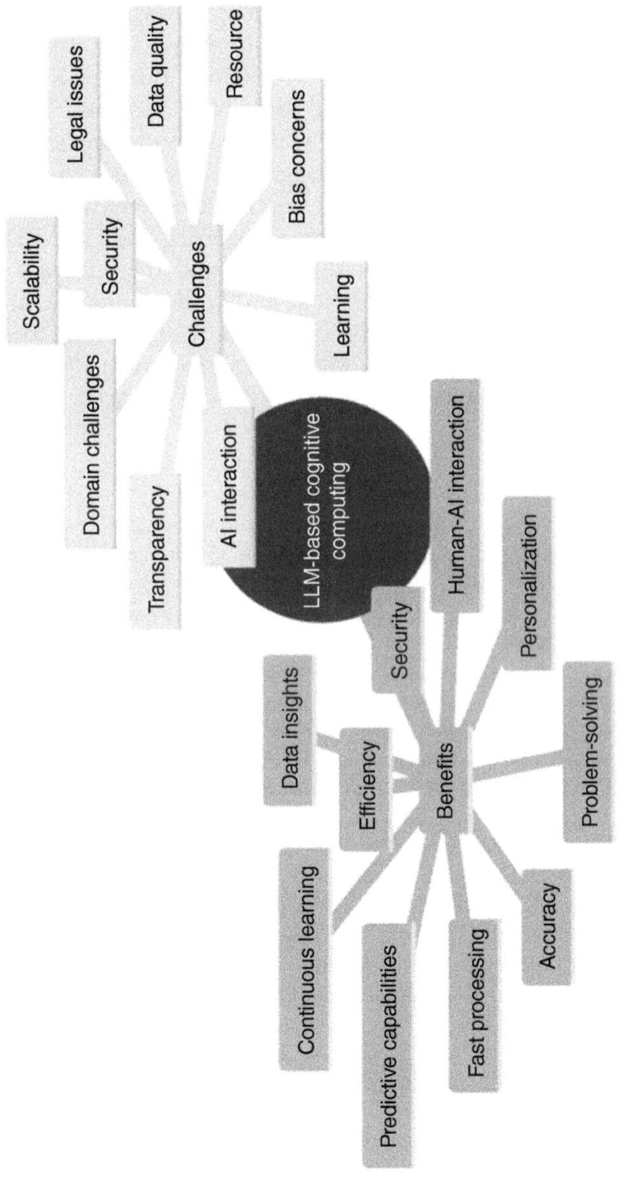

Figure 11.6 Advantages and Disadvantages of LLM-based Cognitive Computing.

11.7 CONCLUSION

Integrating LLMs into cognitive computing frameworks marks a significant leap forward in artificial intelligence. By harnessing the power of LLMs, cognitive computing systems can better process unstructured data, grasp context, and interact more naturally with humans. This study shows how LLMs enhance cognitive systems' performance through advanced NLP, deeper contextual understanding, and interactive learning. The case study on chemo brain rehabilitation highlights the real-world applications of combining LLMs with cognitive computing, demonstrating the potential for personalized and adaptive healthcare interventions. However, integrating LLMs into cognitive computing is not without its challenges. Data biases, the need for more computational resources, and the difficulty of making decision-making processes transparent and understandable are some of these issues. The development of more efficient algorithms, the establishment of strong ethical guidelines, and the improvement of interpretability of cognitive systems should be the focus of future research.

REFERENCES

Achiam, Josh, Steven Adler, Sandhini Agarwal, Lama Ahmad, Ilge Akkaya, Florencia Leoni Aleman, Diogo Almeida et al. (2023). "Gpt-4 technical report." *arXiv preprint arXiv:2303.08774*.

Aghav-Palwe, S., & Gunjal, A. (2021). Introduction to cognitive computing and its various applications. In Manoj Mittal, R. R. Shah and S. Roy (Eds.) *Cognitive computing for human-robot interaction* (pp. 1–18). Academic Press.

Beltagy, I., Peters, M. E., & Cohan, A. (2020). Longformer: The long-document transformer. *arXiv preprint arXiv*:2004.05150, 1-17.

Brown, Tom B., Benjamin Mann, Nick Ryder, Melanie Subbiah, Jared Kaplan, Prafulla Dhariwal, Arvind Neelakantan et al. (2020). "Language models are few-shot learners. arXiv 2020." *arXiv preprint arXiv:2005.14165* 4.

Chen, M., Herrera, F., & Hwang, K. (2018). Cognitive computing: Architecture, technologies and intelligent applications. *IEEE Access, 6*, 19774–19783

Devlin, J., Chang, M. W., Lee, K., & Toutanova, K. (2019). BERT: Pre-training of deep bidirectional transformers for language understanding. In *Proceedings of the 2019 conference of the North American chapter of the association for computational linguistics: human language technologies, volume 1 (long and short papers)*, pp. 4171-4186. 2019.

Dhanalakshmi, R., & Sri Devi, T. (2020). Adaptive cognitive intelligence in analyzing employee feedback using LSTM. *Journal of Intelligent & Fuzzy Systems, 39*(6), 8069–8078.

Dwivedi, Y. K., Kshetri, N., Hughes, L., Slade, E. L., Jeyaraj, A., Kar, A. K., ... & Wright, R. (2023). Opinion Paper:"So what if ChatGPT wrote it?" Multidisciplinary perspectives on opportunities, challenges and implications of generative conversational AI for research, practice and policy. *International journal of information management*, *71*, 102642.

Goodfellow, Ian, Yoshua Bengio, Aaron Courville, and Yoshua Bengio. (2016). *Deep learning*, *1*(2), Cambridge: MIT press.

Hochreiter, S. (1997). Long Short-term Memory. In *Neural Computation MIT-Press*. Carnegie Mellon University.

Hupkes, D., Giulianelli, M., Dankers, V., Artetxe, M., Elazar, Y., Pimentel, T., ... & Jin, Z. (2022). State-of-the-art generalisation research in NLP: a taxonomy and review. *arXiv preprint arXiv:2210.03050.*

Hurwitz, J., Kaufman, M., Bowles, A., Nugent, A., Kobielus, J. G., & Kowolenko, M. D. (2015). *Cognitive computing and big data analytics* (Vol. 288). Wiley.

Khalil, R. A., Jones, E., Babar, M. I., Jan, T., Zafar, M. H., & Alhussain, T. (2019). Speech emotion recognition using deep learning techniques: A review. *IEEE access*, *7*, 117327-117345.

Mikolov, T., I. Sutskever, K. Chen, G. Corrado, and J. Dean (2013). "Efficient estimation of word representations in vector space Tomas." In *IJCAI International Joint Conference on Artificial Intelligence. https://doi.org/10.1162/153244303322533223*. 2013.

Peters, M. E., Neumann, M., Iyyer, M., Gardner, M., Clark, C., Lee, K., & Zettlemoyer, L. (2018). Deep contextualized word representations. *Proceedings of the 2018 Conference of the North American Chapter of the Association for Computational Linguistics: Human Language Technologies*, Volume 1 (Long Papers). pp. 2227–2237.

Pinker, Steven. The language instinct: How the mind creates language. Penguin UK, 2003.

Radford, Alec, Jeffrey Wu, Rewon Child, David Luan, Dario Amodei, and Ilya Sutskever. (2019). "Language models are unsupervised multitask learners." *OpenAI blog*, *1*(8), 9.

Radford, Alec, Karthik Narasimhan, Tim Salimans, and Ilya Sutskever. (2018). "Improving language understanding by generative pre-training.": 3.

Raffel, C., Shazeer, N., Roberts, A., Lee, K., Narang, S., Matena, M., ... & Liu, P. J. (2020). Exploring the limits of transfer learning with a unified text-to-text transformer. *Journal of Machine Learning Research*, *21*(140), 1–67.

Rosenfeld, R., & Clarkson, P. (1997). *Statistical language modeling using the CMU-Cambridge toolkit*. 10.1162/neco.1997.9.8.1735.

Sabarmathi, K. R., & Leelavathi, R. (2019). Application of cognitive computing in healthcare. In *Cognitive Social Mining Applications in Data Analytics and Forensics* (pp. 265–272). IGI Global.

Schuetz, S., & Venkatesh, V. (2020). The rise of human machines: How cognitive computing systems challenge assumptions of user-system interaction. *Journal of the Association for Information Systems*, *21*(2), 460–482.

Shneiderman, Ben. *Designing the user interface: strategies for effective human-computer interaction*. Pearson Education India, 2010.

Turing, A. M. (2007). Computing machinery and intelligence. In *Parsing the Turing test: Philosophical and methodological issues in the quest for the thinking computer* (pp. 23-65). Dordrecht: Springer Netherlands.

Turing, A. M. (2009). Computing machinery and intelligence. In B. J. Copeland (Ed.), *The Essential Turing: Seminal Writings in Computing, Logic, Philosophy, Artificial Intelligence, and Artificial Life* (pp. 441–464). Oxford University Press.

Turing, Alan M. (2009) "Computing machinery and intelligence." Parsing the Turing test: Philosophical and methodological issues in the quest for the thinking computer. Dordrecht: Springer Netherlands, 23-65.

Vaswani, A., Shazeer, N., Parmar, N., Uszkoreit, J., Jones, L., Gomez, A. N., ... & Polosukhin, I. (2017). Attention is all you need. *Advances in Neural Information Processing Systems*, *30*, 5998–6008. https://arxiv.org/abs/1706.03762

Wang, L. (1997). On competitive learning. *IEEE Transactions on neural networks*, *8*(5), 1214–1217.

Wei, Jason, Xuezhi Wang, Dale Schuurmans, Maarten Bosma, Fei Xia, Ed Chi, Quoc V. Le, and Denny Zhou (2022). "Chain-of-thought prompting elicits reasoning in large language models." *Advances in Neural Information Processing Systems* 35, 24824-24837.

Weizenbaum, J. (1966). ELIZA—A computer program for the study of natural language communication between man and machine. *Communications of the ACM*, *9*(1), 36–45.

Zaheer, M., Guruganesh, G., Dubey, K. A., Ainslie, J., Alberti, C., Ontanon, S., ... & Ahmed, A. (2020). Big bird: Transformers for longer sequences. *NeurIPS*. https://arxiv.org/abs/2007.14062

Zhao, W. X., Zhou, K., Li, J., Tang, T., Wang, X., Hou, Y., ... & Wen, J. R. (2023). A survey of large language models. *arXiv preprint arXiv:2303.18223*, *1*(2).

Chapter 12

Electroencephalogram Feature Extraction Methods for Stress Analysis

Radha Shirbhate and Pradip M. Jawandhiya

12.1 INTRODUCTION

In contemporary society, stress has become a widespread mental health concern affecting people across the globe. Those suffering from depression, in particular, often face significant difficulties in managing it. Life becomes extremely difficult, if not impossible. Despair, emptiness, loss, impatience, pessimism, constant tiredness, a sense of isolation, and other like emotions and experiences lead people to withdraw from society and their surroundings over time. Self-harm is a common symptom of many mental health-related or psychological diseases, and a patient may feel compelled to take their own life at some point towards the conclusion of their ordeal. Timely assessment of stress is crucial for protecting an individual's health and well-being. Psychiatrists, psychologists, and clinicians use responses to a set of questions, either attained from the patients or other healthcare studies, to establish if a patient is healthy or clinically depression. If diagnosed with depression, the next step is to determine the severity [1]. There are a number of obstacles to these methods that can make dealing with complex mental health issues difficult, and unfortunately, prone to error. These include the doctors' observations, the patients' reluctance or irritation at having to answer questions, the mental battles that take place between the patients and doctors, the high cost of counselling sessions, and so on. This is why scientists and doctors are continuously on the lookout for better ways to accurately diagnose stress.

An electroencephalogram (EEG), also known as a brain scan, is a non-conventional diagnostic tool. EEG, or electroencephalogram, is a technique that analyses the electrical activity of the brain. It is an excellent tool for learning about brain functions [2]. Brain cells communicate with one another using electric waves. The EEG's signals have been used in the diagnosis of a wide range of illnesses. EEG readings may be an essential biomarker for the diagnosis of stress because feelings of melancholy influence brain activity, and because EEG measurements monitor the electrical activity of the brain as a result of the synaptic connection between the two. Since such

DOI: 10.1201/9781003501220-12

an analysis is dependent on the specialized expertise of the psychiatrist, the graphical interpretation of long-term EEG data is challenging and prone to cognitive mistakes [3]. This makes it difficult to determine whether a patient is experiencing depression or is simply a typical patient. Because EEG is a complex, nonlinear, and time-varying signal, interpreting it visually is time-consuming, taxing, and fraught with the possibility of error.

12.1.1 Problem Statement

At present, a low-cost and straightforward approach to stress detection employing a single-channel device does not exist. In order to solve this issue, we will employ suitable feature extraction and classification methods to develop a human stress identification system for brain signals that operates in real time.

1. **Stress:** The effectiveness of the stress recognition system is limited by an inadequate choice of EEG channels and frequency bands.
2. **Anxiety:** Anxiety not accounted for or anxiety included in the analysis.
3. **Diagnosis:** There isn't yet an easy-to-use, low-cost method for detecting stress using a single-channel sensor.

Effective feature extraction approaches for human stress identification, such as powerful EEG single preparation and substantial psychological examinations (emotional assessment), still need to be studied and improved upon. The major goal of the stress detection framework is to provide a reliable system for extracting and classifying data.

12.2 RELATED WORK

Recent years have seen a surge in interest in the categorization of emotions in terms of brain activity, mostly as a result of the many studies undertaken by scientists from all over the world [4]. Their research revealed that there are numerous techniques for extracting information and classifying EEG signals. Most studies have used numerous machine learning (ML) algorithms tailored to analyze face images, speech signals, and EEG data. While these responses do a good job of outlining the primary components that influence human stress, they barely scratch the surface of the foundation for modern human stress. This chapter highlights the significant breakthroughs of contemporary researchers in mental healthcare and provides insight into the methodologies behind these achievements, particularly from the perspective of EEG signal processing.

The total number of responses from each country indicates that there is a large cross-national variation in the perceived risk of word-related stress, and that this variation cannot be adequately accounted for by socio-economic factors, working conditions, or workplace dissatisfaction. This holds true for each and every one of the national stress indices that have been developed. For each relapse model of country's relative ranking in terms of the amount of stress-related side effects studied. Nearly 80% of people experience stress on the job, and the results of this study point to numerous fascinating and concerning elements of this phenomenon. Statistics show that 60% of workers have quit due to stress on the job. When offered stress management training, more than 90% of employees take advantage of it [6].

Successful implementation of a single-channel stress detection system utilizing the EEG method will ultimately result in the efficient development of a plan of action. Traditional stress detection systems produce inconsistent findings, requiring expensive hardware investments [7]. The EEG method uses digital analysis to determine a person's emotional rhythm. The following section presents a concise review of the most significant research efforts and methodological advancements from the past two decades aimed at identifying and assessing stress.

In 2022, Asghar, Khan, Rizwan, Shorfuzzaman and Mehmood devised a more rapid method for classifying depression [1]. Intrinsic mode functions (IMF) were recovered from the recorded EEG data using intensive Multivariate Empirical Mode Decomposition (iMEMD). To get at the time- and space-related information we needed, we employed a technique called Complex Continuous Wavelet Transform (CCWT). This attribute vector was constructed by integrating the outputs of three independent deep neural networks (DNN) used for attribute mining. We were able to cut down on the attributes' excess bulk by using mutual information and differential entropy. The vector was drastically reduced using the k-means pooling technique. Two standard datasets in the business world were used to evaluate the constructed SA system. Experimental results showed that this framework, despite the difficulty of its development, enhanced classification accuracy and provided a cheap source of real-time SA.

In 2022, Nijhawan, Attigeri and Krishna developed a methodology for identifying stressful social media messages using deep learning [2]. The Bidirectional Encoder Representations from Transformers (BERT) method was used to classify the tone of tweets from a huge dataset. Because of its effectiveness in evaluating textual inputs, the Latent Dirichlet Allocation (LDA) method was applied to a document-wide search for the aim of detecting terms. Tests for depression and stress were given after the post's tone was deduced from its context. Multiple experiments were conducted to examine the efficacy of this depression analysis in improving the detection rate of emotions in social media engagement.

In 2022, Chatterjee and Byun used a stacking ensemble framework to correctly categorize the EEG data [3]. Features were extracted from EEG data after careful analysis. These features were used as input in a stacked ensemble model composed of the Gradient Boosting Classifier (GBC), Random Forest, and Light Gradient Boosting Machine (Light GBM) frameworks, all of which were trained to classify emotions. The predictions of the meta-classifier were then retrieved. The built prediction model outperformed prior state-of-the-art approaches in experimental comparisons of classification accuracy.

In 2022, the cutting edge in CNN (Convolution Neural Network) emotion detection was implemented by Dai, Xi, Li and Wang [4]. More than 4000 data points were used as input for this approach. Properties from the bottom, top, and middle convolution layers were extracted using a feature visualization technique. After that, we verified the scatter plot. The suggested model was found to be more accurate and faster than several existing emotion classification approaches, and additional disentangling of the resulting variables was possible, as evidenced by the experiments.

Muoz and Iglesias implemented a method in 2022 based on textual data from a variety of English-based datasets [5]. Combining the dispersed representations with the lexical features allowed us to fine-tune the categorization procedure. Social, emotional, syntactic, and topical aspects were extracted using the lexicon-based attribute extraction approach. To gather the dispersed representations, we turned to the word embedding technique. The features were supplied into three distinct ML prototypes. F1-score and classification accuracy have both been enhanced by the proposed FastText approach.

In 2022, Sharma, Bohat, Habib, Al-Zoubi, Faris, and Aljarah shortened the EEG input for event-related potentials. The entropy properties were recovered from the EEG data with the help of a Wavelet Transform-based (WT-based) decomposition approach [6]. The approach employed for attribute mining is a stationary one. By employing a support vector machine (SVM), in which hyper-parameters are tuned via a more nuanced heuristic approach, the ESR was performed. The same heuristic method was used to perform weighted attribute fusion as input to the SVM in classification. In this research, we used the Whale Optimization Algorithm (WOA) as our heuristic algorithm of choice, and our heuristically built ESR SVM model outperformed its non-heuristic counterpart in terms of accuracy and consistency [9].

In 2017, Artificial Neural network (ANN) and the filtered Bi-Spectrum (Bi-S) method were incorporated in an ESR system by Purnamasari, Putri Ratna and Kusumoputro [7]. We were able to determine the phase relationship between the EEG's non-linear components by employing the Bi-S technique. After that, we employed a pyramid sample with three dimensions to select information from the Bi-S. This time, a Probabilistic Neural Network (PNN) was used to classify the data [8]. The experimental results showed that the PNN was superior to the baseline in terms of stress recognition performance while requiring significantly less computational effort. Using Bi-S for ESR was shown to be a prudent choice in this investigation.

C. Lin et al. (2018) investigated the correlation between keyboard strikes and occupational stress. Short- and long-term assessments of the participants revealed that many of them actually suffered from a wide range of medical issues [10]. Several methods, spanning a range of approaches, have been reported for use in establishing pressure. A number of studies, including those by C. Viegas and Maxion (2018), have focused on participants' facial expressions as a means of detecting stress while others have looked for signs of stress through the temperature of the fingers, human gestures, and blinking [11].

12.3 PROPOSED METHOD

Data gathering is the process of obtaining, analyzing, and interpreting various types of information from various sources is what data collection is all about. Research often involves collecting data in order to learn all there is to know about a topic and to set the stage for future decisions [12–16]. It helps healthcare systems priorities patients, tailor treatments to each individual, streamline treatment processes, enhance doctor-patient communication, and ultimately improve health outcomes. The electrical activity of a system, component, or part can be evaluated using electroencephalogram (EEG), a comprehensive set of non-invasive data collection and analysis methods [17–20]. Using the subject's stress and non-stress states in EEG has been shown to be a viable and dependable method. Before identifying emotions, this method separates bands, selects bands, extracts features, and processes the data. The EEG data is captured at a frequency of 512 Hz. This dataset makes use of the pre-processed EEG data collection that underwent down sampling (128 Hz) and filtering (4 to 45 Hz). Following the global 10-20 system, the 32 electrodes are placed as follows: Fp1, AF3, F3, F7, FC5,FC1, C3, T7, CP5, CP1, P3, P7, PO3, O1, Oz, Pz, Fp2, AF4, Fz, F4, F8, FC6, FC2, Cz, CP6, CP2, P4, P8, PO4, and O2. Electrode interaction with neural tissue allows for substantial information about neural signals and improves diagnosis and treatment of neurological diseases.

The fundamental units of the brain responsible for generating brain waves are distinguished by their band **separation and selective activity**. Neurons are single cells that communicate with one another via electrical fluctuations. A test known, now introduced as an electroencephalogram or EEG shows brain wave patterns. The frequency of a brain wave is measured in hertz (Hz), which represents cycles per second. The brain waves in question span multiple frequency ranges. Since the development of EEG, researchers have been interested in the many different types of brain rhythms and how they relate to different diseases and abilities [21]. Exercises, neural activity, and the fundamental frequency ranges of EEGs are used to distinguish normal and abnormal brain rhythms, reflecting various mental states, cognitive abilities, and neurological conditions There are several popular uses for EEG signals, including seizure detection and prediction, motor imagery classification, emotional task classification, sleep state classification, and drug effects diagnosis. Given the vast volume of EEG data, careful selection of frequency bands is crucial, with their significance varying according to the specific application.

Band selection methods in EEG signal processing serve multiple critical objectives:

- Reducing the computational complexity of any processing tasks performed on EEG signals by picking the band and extracting desired features is the fundamental goal of the band selection method. In other words, fundamentally reducing computational complexity through strategic frequency band selection and targeted feature extraction.
- In order to improve performance, minimize the amount of overfitting that can occur and lead to wasteful band usage. In short, enhancing overall system performance while mitigating overfitting and inefficient spectral resource utilization.
- To shorten setup times in certain applications. Put simply, streamlining implementation timelines for practical applications.

12.3.1 Feature Extraction and Selection

There may be a number of benefits to using feature selection methods. Minimizing the amount of time spent training, using, storing, and measuring are just a few of these. The primary objective of feature selection techniques is to improve classification accuracy through the strategic identification and selection of the most informative feature subsets. The existence of uncorrelated features reduces the classifiers' performance. Various feature extraction methodologies are utilized to select discriminative features that enable accurate classification of six distinct emotional states. These include RER (Relative Energy Ratio), KDE (Kernel Density Estimation), Hjorth (Hjorth Parameters), ELC (Energy Level Coefficient), and BFCC (Bark Frequency Cepstral Coefficients).

12.3.2 Emotion Detection

The methods used in emotion detection with EEG are diverse, but the ultimate aims of the study are consistent. Through the application of several analytical techniques, our goal is to discover suitable features for emotion identification. The next step is to use an improved model for categorization with the aim of improving emotion recognition accuracy. Additionally, we aim to establish a strong physiological foundation for studies utilizing electroencephalograms (EEGs) to detect emotions by identifying the most relevant frequency bands and brain areas associated with these activities [22].

12.3.3 Stress Recognition

The ability to identify and manage one's own emotions is more valuable than ever in today's world. Even among psychologists, there is a lack of consensus regarding the number and meaning of distinct emotions [23]. Brain and psycho physiological signals have received a lot of attention from researchers in their own right, but their interplay has received surprisingly little consideration. When people are not dealing with stress, their mental states are typified by feelings of contentment, joy, and tranquilly. When people are under stress, on the other hand, negative emotions like wrath, fear, and despair are more common. An individual's emotional health is a good indicator of how much stress they are under. A person is said to be unstressed when they are feeling joyful and hopeful, and stressed when they are feeling angry and negative [24–26].

12.3.4 Classification Techniques

Even after obtaining the desired features through the feature extraction process, researchers must still address the challenge of emotion classification. This task will be performed by a classifier. By dividing data into numerous classes, a classifier can find the link between qualities and emotional state. The issue of human stress detection using classification algorithms has numerous approaches, each of which compares and contrasts the effectiveness of various feature extraction methods

Using a synthesis of EEG data recorded during exposure to neutral, negative, and beneficial auditory stimuli, the authors of this research provide a multi modal stress detection model. The following are the pillars upon which the synthesis rests, as seen in Figure 12.1.

EEG studies have revealed distinct neurophysiological patterns in individuals with depression that significantly alter their emotional processing capabilities.

1. Depressed people, according to EEG data, have a reduced subjective sense of positive emotional stimuli (lower positive sensations).
2. People who suffer from depression have a heightened emotional reactivity (bad emotions are enhanced) and a heightened emphasis on unpleasant feelings in response to unfavorable expressive cues.

The obtained structures from two modalities using two different feature extraction methods (positive audio stimulus and negative audio stimulus,

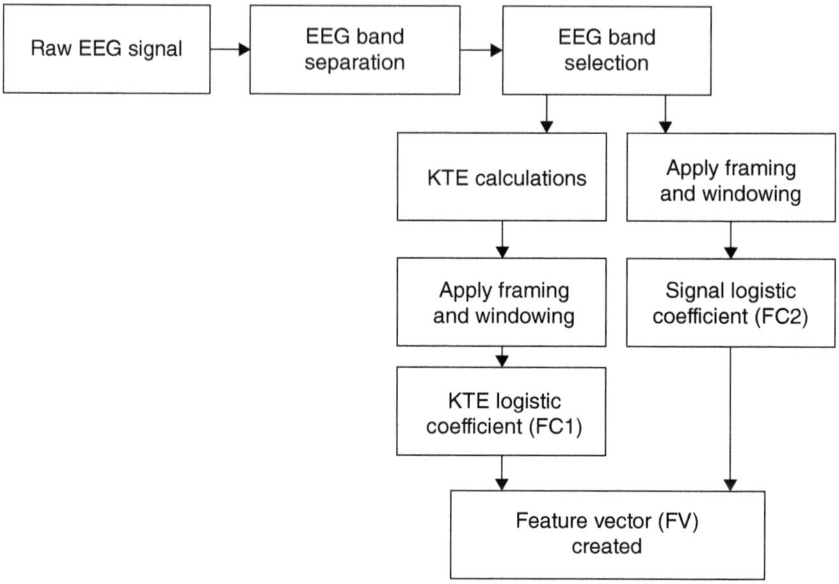

Figure 12.1 Algorithm for Feature Extraction.

respectively) are incorrect if there are any discrete differences between them. Given the challenges, it is possible that mixing traits from multiple modalities can make up for the lack of individual modalities. Feature-level integration of EEG data is performed in this study.

12.4 METHODOLOGY

The outline depicts a computerized system that uses EEG signals and sentiment analysis to detect signs of stress in humans and help them cope with the condition. Multiple actions and mental states can be linked to the unique electrical frequencies revealed by EEG analysis.

Figure 12.2 of the EEG analysis demonstrates the resulting wide range of abundance responses to both extrinsic and endogenous stimuli. The methodology for distinguishing between stressed and unstressed states involves a systematic pipeline of pre-processing, EEG band separation, band selection, feature extraction, and emotion detection with sentiment analysis, as illustrated in the figure. Figure 12.2 also illustrates the complete workflow of the proposed methodology.

The purpose of study to develop a user-independent system for detecting stress in humans using EEG data. Sentiment analysis for human emotion recognition and stress screening; finding the best and most efficient features.

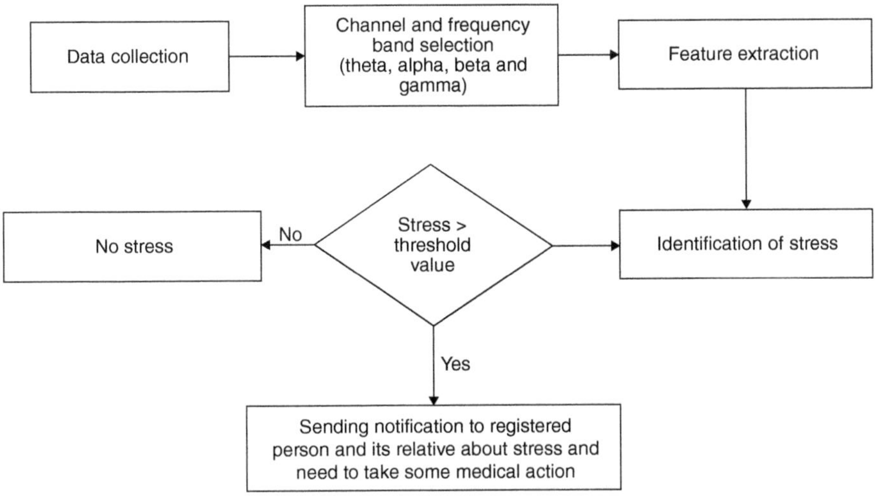

Figure 12.2 Human Stress detection Approach.

What follows is a summary of the state of the art in feature extraction. Significant outcomes can be achieved with BCI applications using PCA, ICA (Independent Component Analysis), and EMD. Using SVM in the software development process has various advantages: When it comes to stress and pressure detection employing EEG signals, PCA (which, to remind the reader, stands for principal component analysis) exhibits significant generalization qualities; yet, its influence is subject to the curse-of-dimensional if the signals are negative. These advantages may be attained by utilizing EEG signals.

Pre-processing outcomes on selected brain signals. One of the current approaches for assessing mental state is clearly superior when compared to others; this method makes use of feature extraction and influential feature selection. Among the many ways these features are described are their spectral and statistical qualities as well as their band frequency cepstral coefficient properties. All three of these characteristics of the EEG signal are crucial for feature extraction. The proposed method of employing band frequency cepstral coefficients allows for the extraction of features crucial to stress identification. In terms of emotion categorization accuracy, experimental results show that the proposed BFCC and ELC algorithms beat state-of-the-art algorithms such as KDE, RER, and Hjorth parameters. When coping with states of mental illness, unexpected disturbances of brain waves occur, requiring substantial signal processing demands for abnormal state diagnosis [4]. For this reason, the interpretation of stress determination by influential factors outperforms competing ML methods on the benchmark

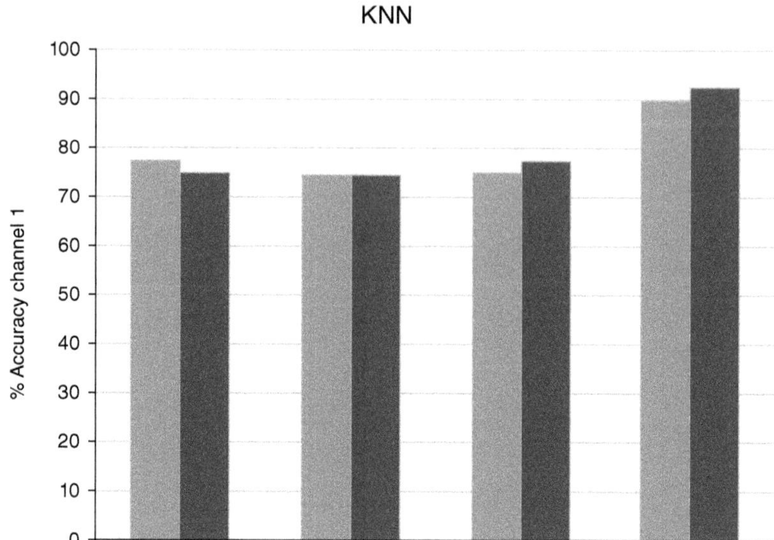

	KDE % Accuracy happy n angry	RER % Accuracy happy n angry	ELC % Accuracy happy n angry	BFCC % Accuracy happy n angry
1 Alpha	77.5	74.5	75	90
1 Beta	75	74.5	77.5	92.5

Figure 12.3 Performance Analysis of KNN.

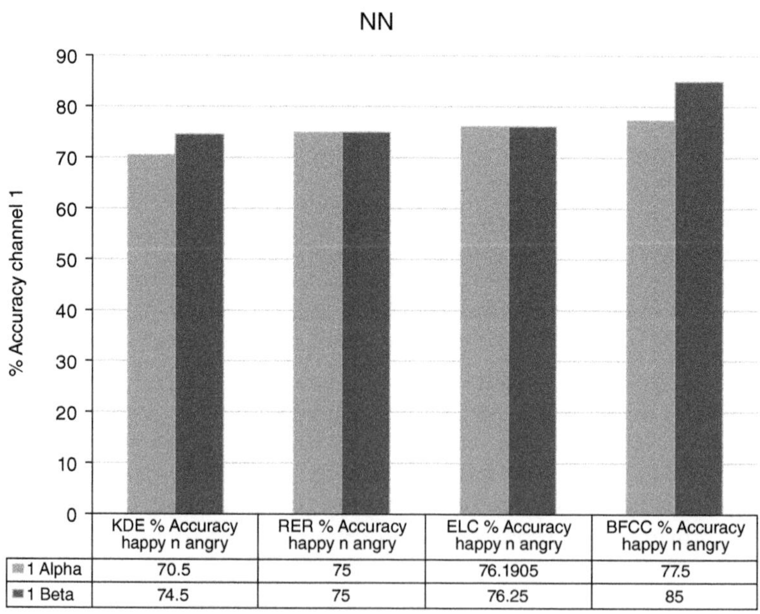

	KDE % Accuracy happy n angry	RER % Accuracy happy n angry	ELC % Accuracy happy n angry	BFCC % Accuracy happy n angry
1 Alpha	70.5	75	76.1905	77.5
1 Beta	74.5	75	76.25	85

Figure 12.4 Performance Analysis of Neural Network.

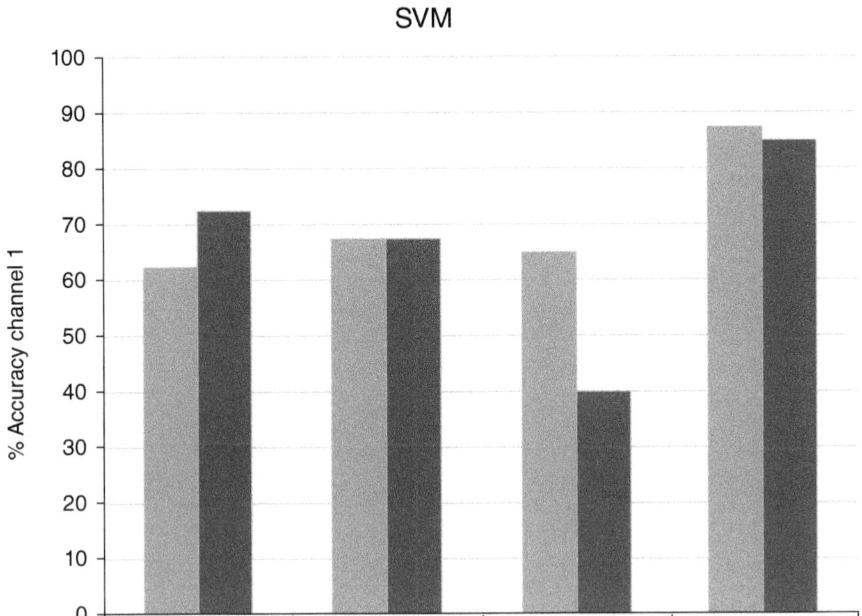

	KDE % Accuracy happy n angry	RER % Accuracy happy n angry	ELC % Accuracy happy n angry	BFCC % Accuracy happy n angry
1 Alpha	62.5	67.5	65	87.5
1 Beta	72.5	67.5	40	85

Figure 12.5 Performance Analysis of Support Vector Machine.

DEAP datasets in terms of prediction rate. Using an electrode insertion technique ranging from 10 to 20, the 32-channel datasets was generated.

The NeuroSky single-channel EEG device is used by an automated stress identification system to track and analyses stress levels in real-time. At the outset, the device gathers data by recording the user's EEG. In order to isolate the alpha, beta, and gamma bands—which are related with stress—from the rest of the signal, the preprocessing procedures involve noise reduction and signal filtering. The filtered signals are then subjected to feature extraction techniques, which include statistical measurements (mean, variance) and power spectral density (PSD). The collected features are used to categories stress levels using ML methods such as Random Forest and SVM. When users' stress levels rise above certain thresholds, the system's auto notification function sends them a notification. Analyzing the system's performance entails testing its precision, recall, and discriminatory power on gathered datasets via cross validation methods. In spite of the fact that the NeuroSky is only a single-channel EEG, the results show that it is

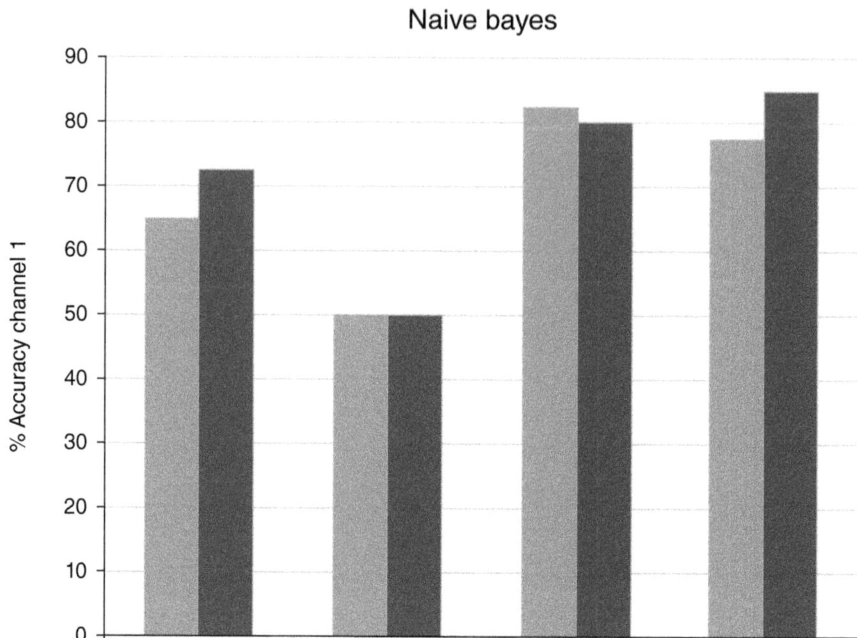

	KDE % Accuracy happy n angry	RER % Accuracy happy n angry	ELC % Accuracy happy n angry	BFCC % Accuracy happy n angry
1 Alpha	65	50	82.5	77.5
1 Beta	72.5	50	80	85

Figure 12.6 Performance analysis of NB.

capable of accurate stress detection (with rates of 75% to 85%). In practical settings, this method shows promise for affordable, portable stress monitoring solutions. Performance analysis for automated stress detection system using various algorithms and accuracy shown in Figure 12.3–12.7 respectively.

12.5 CONCLUSION

This is the first study to use EEG signal data and its geometrical properties to compare healthy and depressed participants. This research is the first to compare and contrast the ways in which optimization strategies for differentiating between healthy and depressed EEG data are presented. Out of many feature choosing procedures and classifiers, we found that SVM classifiers are the most efficient at compressing feature vector displays and identifying depression. We compared nine participants without depression to nine with depression to see if they had the specified profile. With an

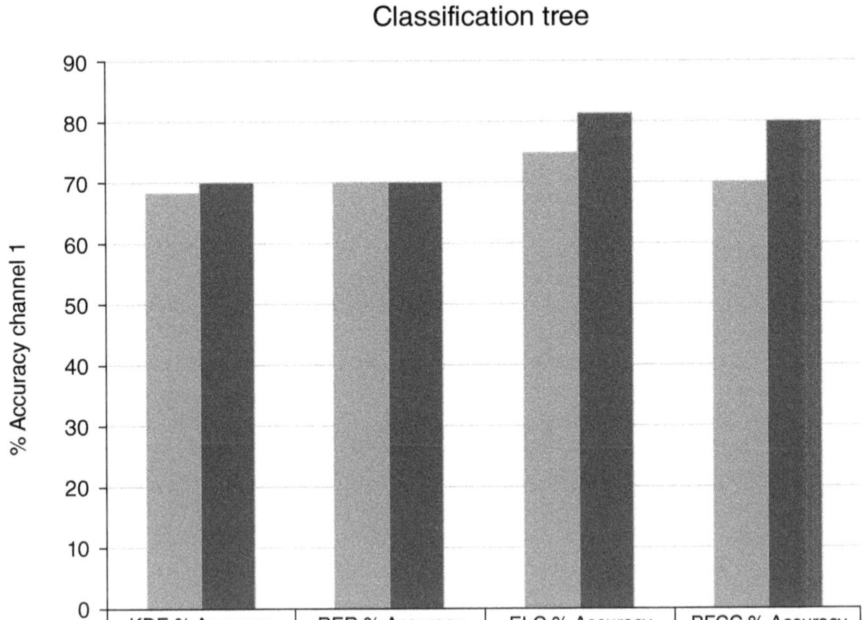

	KDE % Accuracy happy n angry	RER % Accuracy happy n angry	ELC % Accuracy happy n angry	BFCC % Accuracy happy n angry
1 Alpha	68.5	70	75	70
1 Beta	70	70	81.5	82.5

Figure 12.7 Performance analysis of CT.

accuracy of 98%, the proposed approach correctly identified normal and depressed EEG signals. Our proposed computer-assisted screening for stress proved clinically valid and efficient.

12.6 FUTURE SCOPE

There will soon be more research looking at the stress-relationship between EEG signal analysis and a variety of human emotions. Furthermore, diagnostics-based clinical problems can be utilized by biomedical applications. The Integration of EEG data with other physiological signals such as heart rate and skin conductance used for a more comprehensive analysis. Development of adaptive systems that continuously learn and update stress models based on individual variations and changing stress patterns. Personalized stress profiles are created by considering an individual's baseline EEG patterns and incorporating contextual information. The research specifies advanced feature extraction methods, including deep learning-based feature learning approaches. As we know conducting large-scale studies to validate and generalize the effectiveness of EEG-based stress analysis across diverse populations.

In conclusion, EEG-based stress analysis using ML holds great potential for various applications, from mental health monitoring to workplace stress management. Continuous research, technological advancements, and interdisciplinary collaborations are essential for further refining these approaches and addressing real-world challenges.

REFERENCES

[1] Muhammad Adeel Asghar, Muhammad Jamil Khan, Muhammad Rizwan, Mohammad Shorfuzzaman and Raja Majid Mehmood, "AI inspired EEG-based spatial feature selection method using multivariate empirical mode decomposition for emotion classification", *Multimedia Systems*, vol. 28, pp. 1275–1288, 2022.

[2] Tanya Nijhawan, Girija Attigeri and T. Anantha Krishna, "Stress detection using natural language processing and machine learning over social interactions," *Journal of Big Data*, vol. 9, pp. 33, 2022.

[3] Subhajit Chatterjee and Yung-Cheol Byun, "EEG-based emotion classification using stacking ensemble approach," *Sensors*, vol. 22, pp. 8550, 2022.

[4] Jinxiao Dai, Xugang Xi, Ge Li and Ting Wang, "EEG-based emotion classification using improved cross-connected convolutional neural network," *Brain Sciences*, vol. 12, pp. 977, 2022.

[5] Sergio Muñoz, and Carlos A. Iglesias, "A text classification approach to detect psychological stress combining a lexicon-based feature framework with distributional representations," *Information Processing & Management*, vol. 59, no. 5, pp. 103011, September 2022.

[6] Lakhan Dev Sharma, Vijay Kumar Bohat, Maria Habib, Ala' M. Al-Zoubi, Hossam Faris, and Ibrahim Aljarah, "Evolutionary inspired approach for mental stress detection using EEG signal," *Expert Systems with Applications*, vol. 197, pp. 116634, 1 July 2022.

[7] Prima Dewi Purnamasari, Anak Agung Putri Ratna and Benyamin Kusumoputro, "Development of filtered bispectrum for EEG signal feature extraction in automatic emotion recognition using artificial neural networks," *Algorithms*, vol. 10, pp. 63, 2017.

[8] Bishwajit Roy, Lokesh Malviya, Radhikesh Kumar, Sandip Mal, Amrendra Kumar, Tanmay Bhowmik and Jong Wan Hu, "Hybrid deep learning approach for stress detection using decomposed EEG signals," *Diagnostics*, vol. 13, no. 11, pp. 1936, 2023.

[9] J. -L. Wu, Y. He, L. -C. Yu and K. R. Lai, "Identifying emotion labels from psychiatric social texts using a Bi-directional LSTM-CNN model," *IEEE Access*, vol. 8, pp. 66638–66646, 2020.

[10] Luis G. Hernández, Oscar Martinez Mozos, José M. Ferrández and Javier M. Antelis," EEG-based detection of braking intention under different car driving conditions," *Frontiers in Neuroinformatics*, vol. 12, May 2018. Article 29. https://doi.org/10.3389/fninf.2018.00029

[11] Khalid Masood and mohammed a. Alghamdi, "Modeling mental stress using a deep learning framework," *IEEE Access,* vol. 7, pp. 68446–68454, 2019.

[12] Tianqi Fan, Sen Qiu, Zhelong Wang, Hongyu Zhao, Junhan Jiang, Yongzhen Wang, Junnan Xu, Tao Sun, and Nan Jiang, "A new deep convolutional neural network incorporating attentional mechanisms for ECG emotion recognition," *Computers in Biology and Medicine*, vol. 159, pp. 106938, June 2023.

[13] C. Viegas and R. Maxion, "Towards independent stress detection: A dependent model using facial action units," *2018 International Conference on Content-Based Multimedia Indexing*, pp. 1–6, 2018.

[14] S. Koldijk and M. A. Neerincx, "Detecting work stress in offices by combiningunobtrusive sensors," vol. 3045, no. c, 2016.

[15] K. Shahzabeen, A. Wahab, H. A. Majid, and B. Crüts, "Analyzing brain activity in understanding cultural and language interaction for depression and anxiety," *Procedia – Social and Behavioral Sciences*, vol. 27, no. Pacling, pp. 299–305, 2011.

[16] M. Khezri, M. Firoozabadi, and A. Reza, "Reliable emotion recognition system based on dynamic adaptive fusion of forehead biopotentials," *Computational Methods and Programs in Biomedicine*, vol. 122, no. 2, pp. 149–164, 2015.

[17] B. S. Zheng, M. Murugappan, S. Yaacob, and S. Murugappan, "Human emotional stress analysis through time domain electromyogram features," *Proceedings of the IEEE Symposium on Computer Applications and Industrial Electronics (ISCAIE)*, pp. 172–177, 2013.

[18] L. U. Z. Santamaria-granados, M. Munoz-organero, G. Ramirez-gonzález, E. Abdulhay, and N. Arunkumar, "Using deep convolutional neural network for emotion detection on a physiological signals dataset (AMIGOS)," *IEEE Access*, 7, 2019

[19] R. M. Mehmood, R. Du, and H. Y. O. J. Lee, "Special section on advances of multisensory services and optimal feature selection and deep learning ensembles method for emotion recognition from Human Brain EEG Sensors," *IEEE Access* vol. 5, pp. 14797–14806, 2017.

[20] Sarah N. Abdulkader, Ayman Atia, and Mostafa-Sami M. Mostafa.. "Brain Computer Interfacing: Applications and Challenges." *Egyptian Informatics Journal,* vol. 16, no. 2, pp. 213–30, 2015. https://doi.org/10.1016/J.EIJ.2015.06.002

[21] Anita Patil, Ashish Panat, and Supriya Ambadas Ragade. 2015. "Classification of Human Emotions from Electroencephalogram Using Support Vector Machine," . In Proceedings of the International Conference on Advances in Computing, Communications and Informatics (ICACCI) pp. 404–408.

[22] Chin-teng Lin, Jung-tai King, Jhe-wei Fan, Abhishek Appaji, and Mukesh Prasad. "The influence of acute stress on brain dynamics during task switching activities," pp. 3249–3255, 2018.

[23] A Aidatul , M P Paulraj, Sazali Yaacob, and N F Mohamad Nasir. "Automated system for stress evaluation based on EEG signal: A prospective

review," *IEEE 7th International Colloquium on Signal Processing and its Applications,* pp. 167–171, 2011.

[24] Jinyu Li, Haoqiang Hua, Zhihui Xu, Lin Shu, Xiangmin Xu, Feng Kuang, and Shibin Wu, "Cross-subject EEG emotion recognition combined with connectivity features and meta-transfer learning," *Computers in Biology and Medicine*, vol. 145, pp. 105519, June 2022.

[25] Prasannavenkatesan Theerthagiri, "Stress emotion recognition with discrepancy reduction using transfer learning," *Multimedia Tools and Applications*, vol. 82, pp. 5949–5963, 2023.

[26] Hongyuan Xuan Dong Cui, Jing Liu, Guanghua Gu & Xiaoli Li, "Emotion recognition on EEG signal using ResNeXt attention 2D-3D convolution neural networks," *Neural Processing Letters*, vol. 55, No. 5, pp. 5943–5957, 2022.

Index

Note: Page numbers in *italic* denote figures and in **bold** denote tables.

For Product Safety Concerns and Information please contact our EU
representative GPSR@taylorandfrancis.com
Taylor & Francis Verlag GmbH, Kaufingerstraße 24, 80331 München, Germany

www.ingramcontent.com/pod-product-compliance
Ingram Content Group UK Ltd.
Pitfield, Milton Keynes, MK11 3LW, UK
UKHW022220190926
475702UK00003B/455

* 9 7 8 1 0 3 2 7 9 5 3 3 1 *